Geometry Snacks

Ed Southall and Vincent Pantaloni

About the Authors

Ed Southall is a teacher trainer at Huddersfield University in the UK, and the author of *Yes But Why? Teaching for Understanding in Mathematics*. Ed has taught mathematics for 14 years in secondary schools both in the UK and the Middle East. He has developed a large following online as @solvemymaths on twitter where he regularly posts mathematical puzzles.

Vincent Pantaloni is a French mathematics high school teacher since 2000. He teaches in Orléans in French but also in English to the European and International Sections. With these students he has developed a teaching based on cooperative problem solving. He believes that searching and sharing strategies for complex problems in teamwork helps students in building a stronger mathematical knowledge. He is also a teacher trainer specialised in the use of new technologies and flipped classroom and a member of different research groups. He posts about mathematics on twitter @panlepan and in French on his website Mathzani.

ISBN (book): 978-1-911093-70-1
ISBN (ebook): 978-1-911093-71-8

Printed and designed in the UK

Published by Tarquin
Suite 74, 17 Holywell Hill
St Albans AL1 1DT
United Kingdom

info@tarquingroup.com
www.tarquingroup.com

Introduction

Geometry Snacks is a mathematical puzzle book filled with geometrical figures and questions designed to challenge, confuse and ultimately enlighten enthusiasts of all ages. Each puzzle is carefully designed to draw out interesting phenomena and relationships between the areas and dimensions of various shapes. Furthermore, unlike most puzzle books, the authors offer multiple approaches to solutions so that once a puzzle is solved, there are further surprises, insights and challenges to be had.

As a teaching tool, Geometry Snacks enables teachers to promote deep thinking and debate over how to solve geometry puzzles. Each figure is simple, but often deceptively tricky to solve – allowing for great classroom discussions about ways in which to approach them. By offering numerous solution approaches, the book also acts as a tool to help encourage creativity and develop a variety of strategies to chip away at problems that often seem to have no obvious way in.

Contents

Publisher's Note: About the Tarquin eReader

Geometry Snacks is a wonderful addition to Tarquin's range of puzzle, activity and enriching books for lovers of mathematics.

We are delighted to provide a version of the book that is ideal for teachers and those curious to explore the Snacks and their solutions using demonstrations in Geogebra. Using the new free Tarquin eReader, where appropriate Snacks are explored with animations embedded into the ebook. So you can choose whether to read on or click through to find out more.

Perfect for classroom use on whiteboard or computer, as well offering additional insights into the problems. Available early 2018.

For more about this, or to explore our large range of books, posters and resources for all ages go to www.tarquingroup.com or contact us at info@tarquingroup.com or @tarquingroup on Twitter.

What Fraction is Shaded?

The aim of these puzzles is to identify the shaded portion of the whole figure as a fraction of its total area. All of these puzzles are created using regular shapes. Any additional information required to solve each puzzle is indicated in the text below the figure.

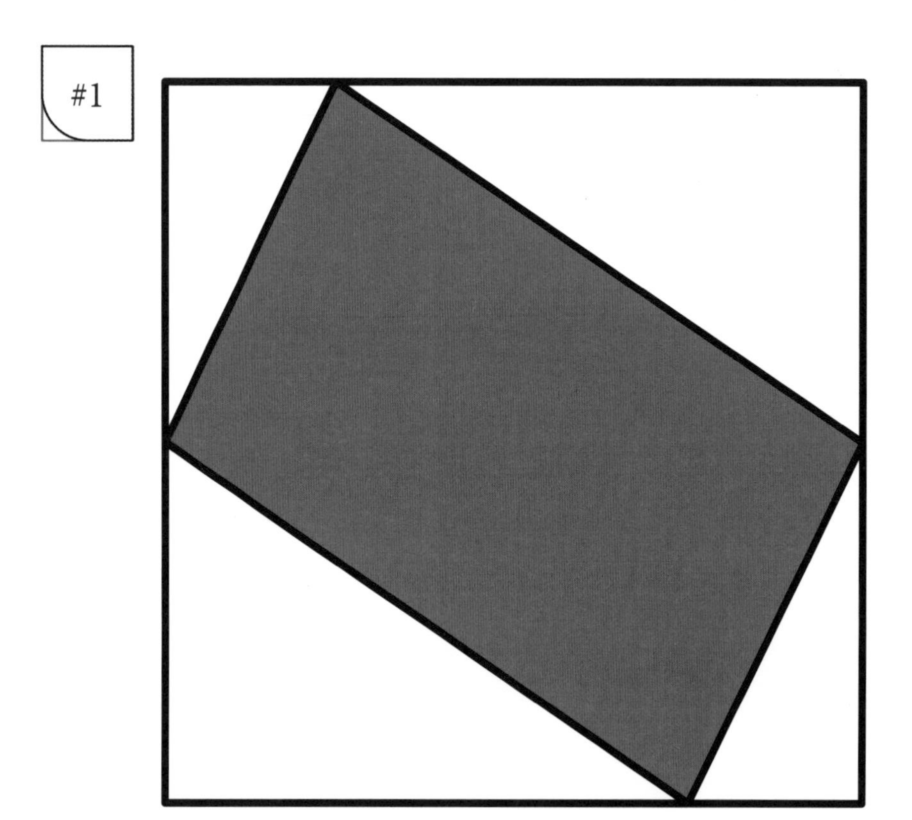

A parallelogram is inscribed in a square such that two vertices are at the midpoints of opposite sides as shown.

What fraction of the square is shaded?

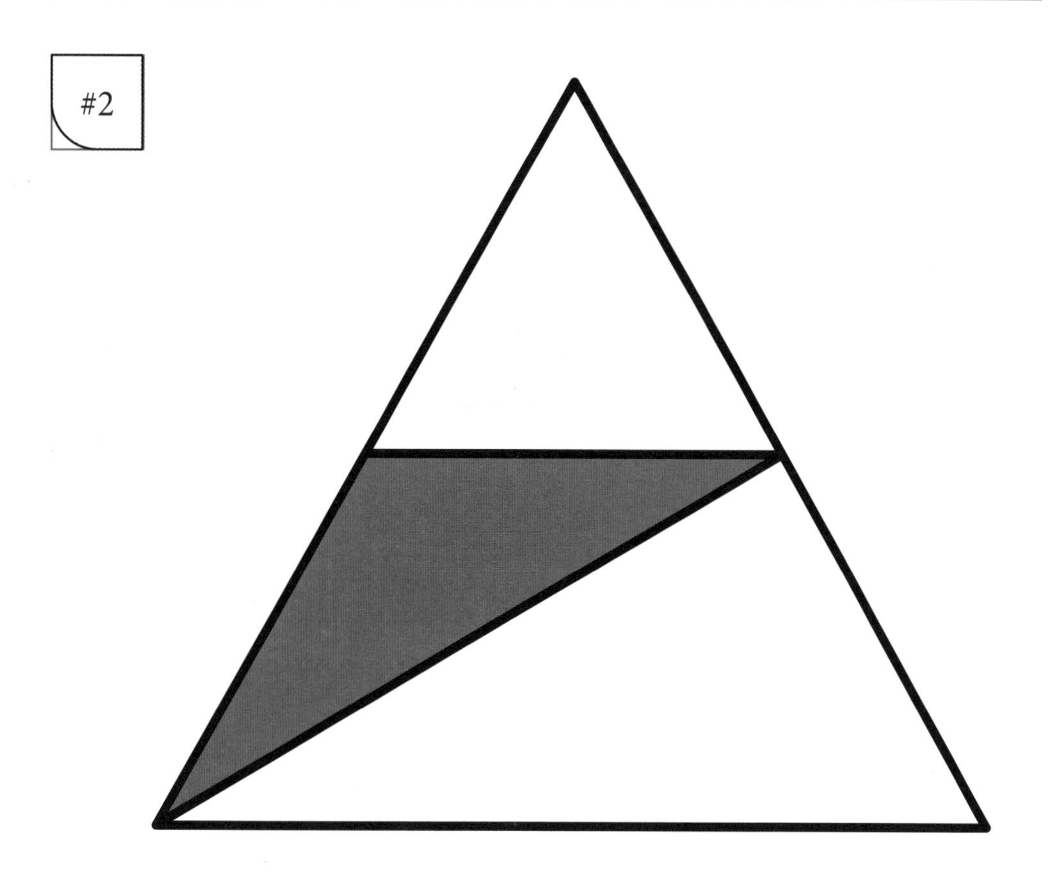

The shaded triangle is constructed using the midpoints and a vertex of an equilateral triangle as shown.

What fraction of the equilateral triangle is shaded?

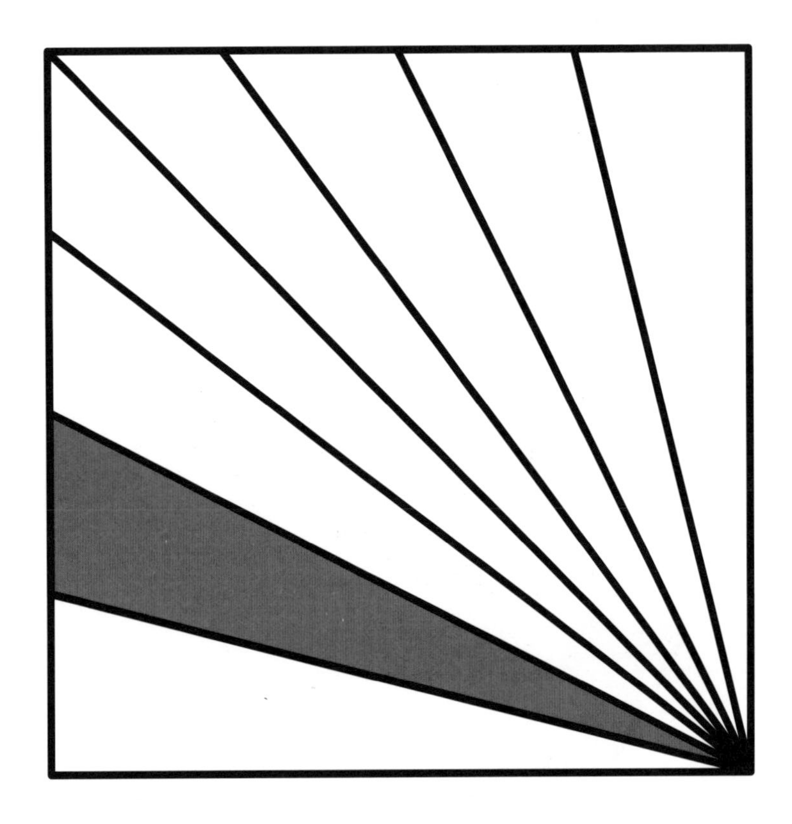

Equally spaced points across two sides of a square form triangles with one shared vertex as shown.

What fraction of the square is shaded?

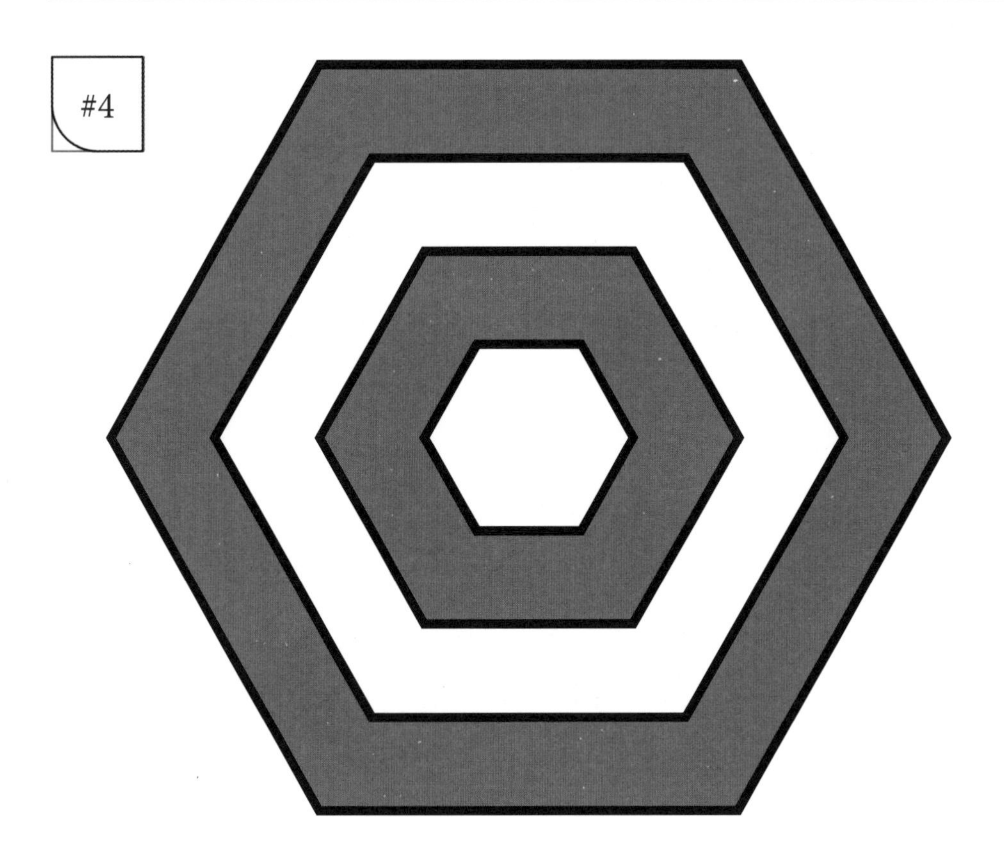

Four concentric regular hexagons are equally spaced apart from the centre point as shown.

What fraction of the largest regular hexagon is shaded?

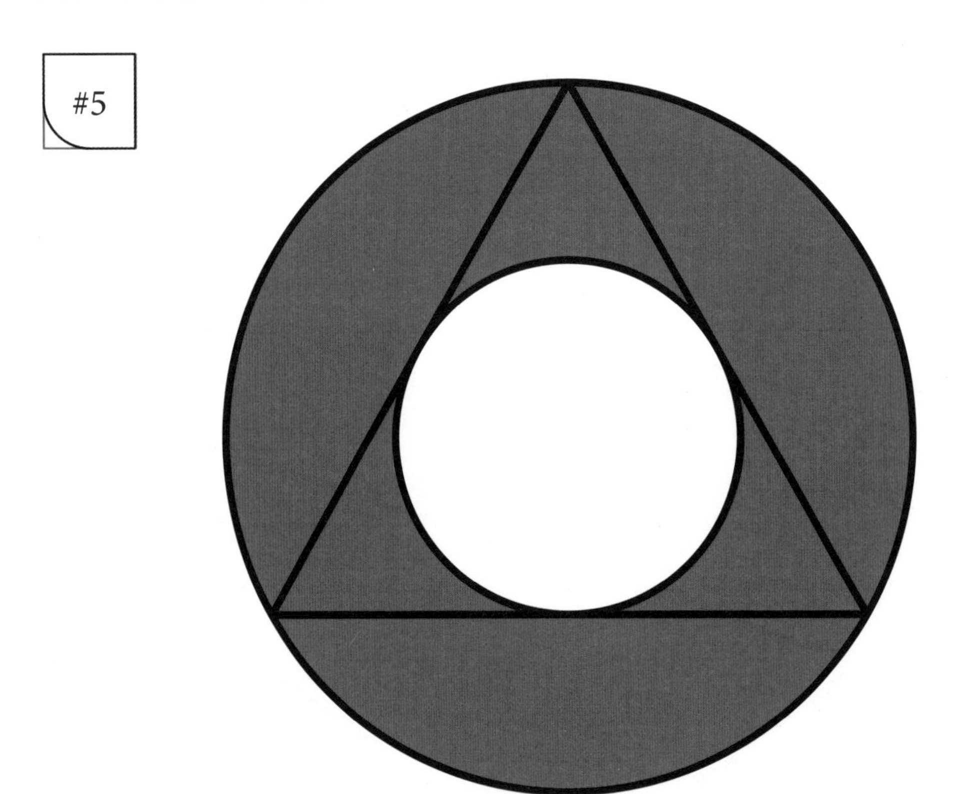

A circumscribed equilateral triangle has an incircle as shown.

What fraction of the larger circle is shaded?

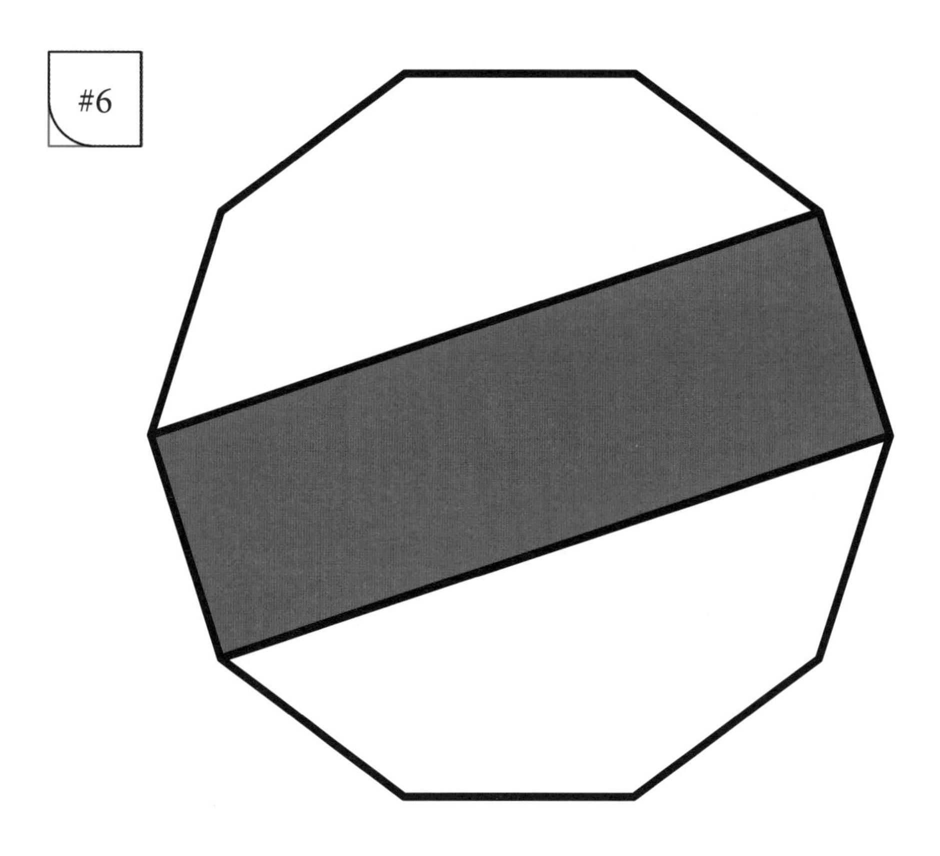

A rectangle is inscribed in a regular decagon as shown.

What fraction of the decagon is shaded?

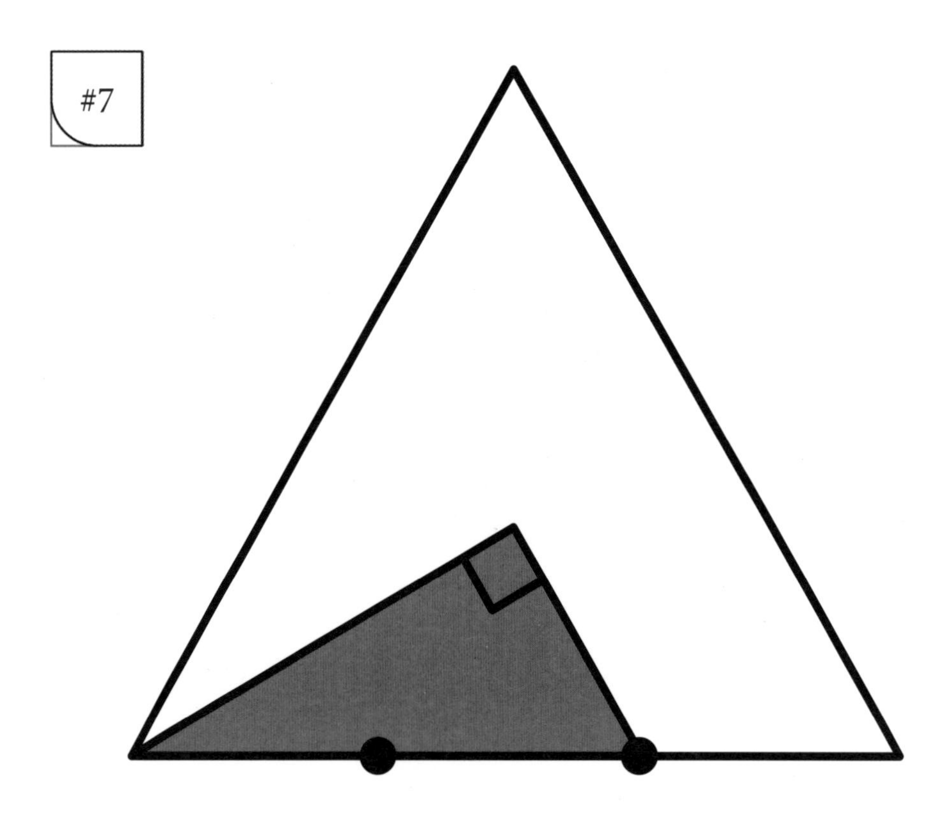

A right angled triangle is constructed within an equilateral triangle using the trisection of a side as shown.

What fraction of the equilateral triangle is shaded?

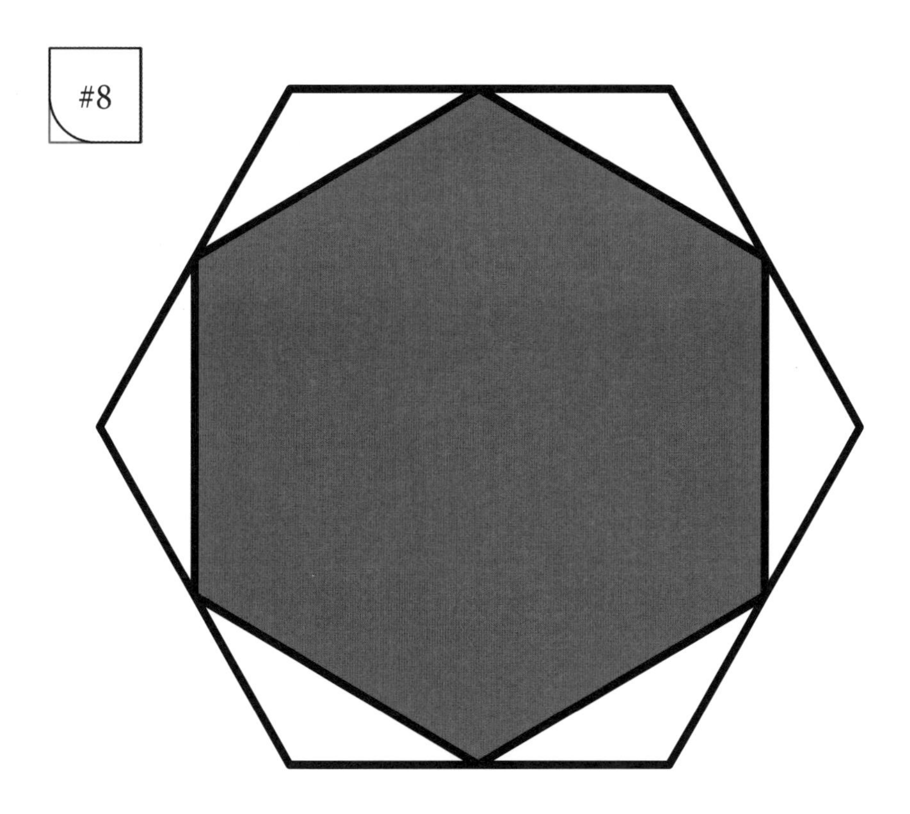

A regular hexagon is constructed using the midpoints of a similar hexagon.

What fraction of the larger hexagon is shaded?

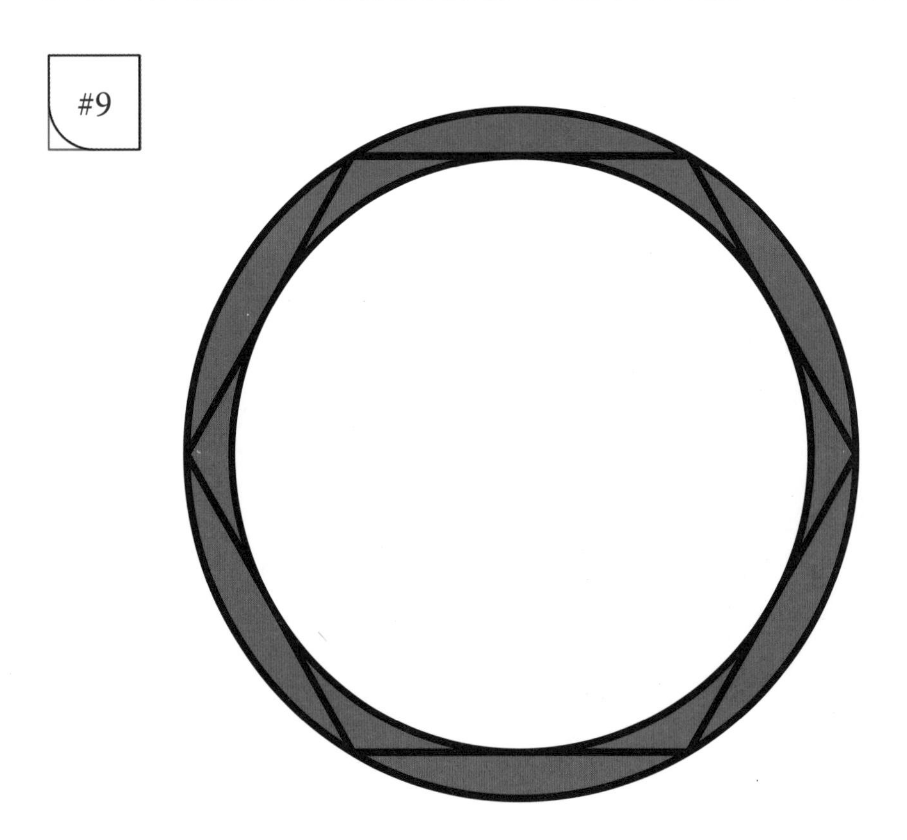

An incircle and circumcircle of a regular hexagon are constructed as shown.

What fraction of the larger circle is shaded?

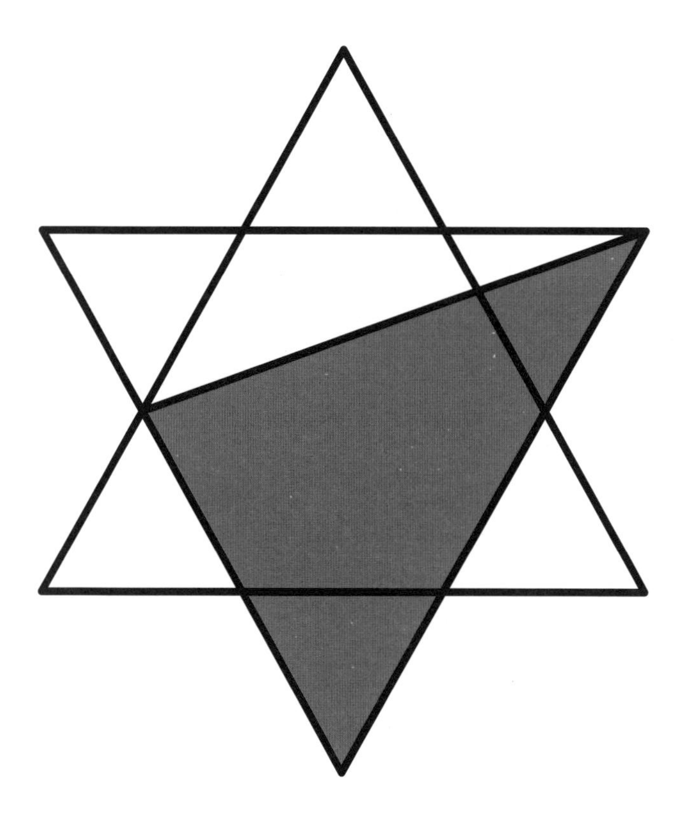

A regular six-pointed star is constructed using two equilateral triangles.

What fraction of the star is shaded?

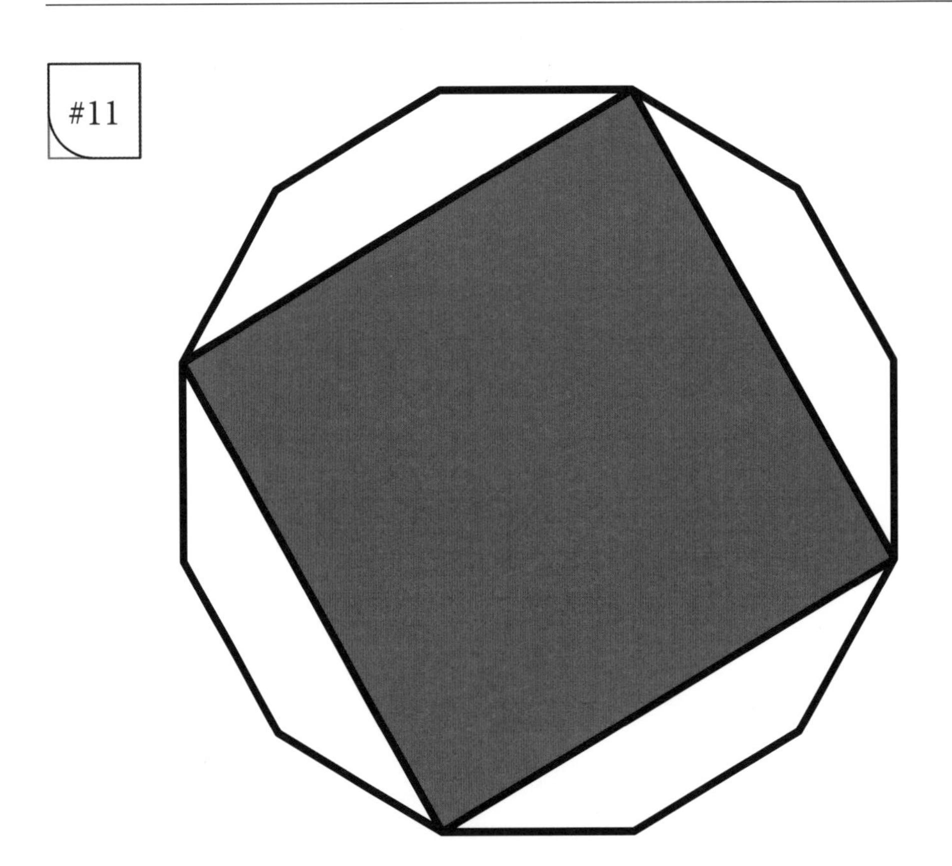

A square is constructed using vertices of a regular dodecagon as shown.

What fraction of the dodecagon is shaded?

#12

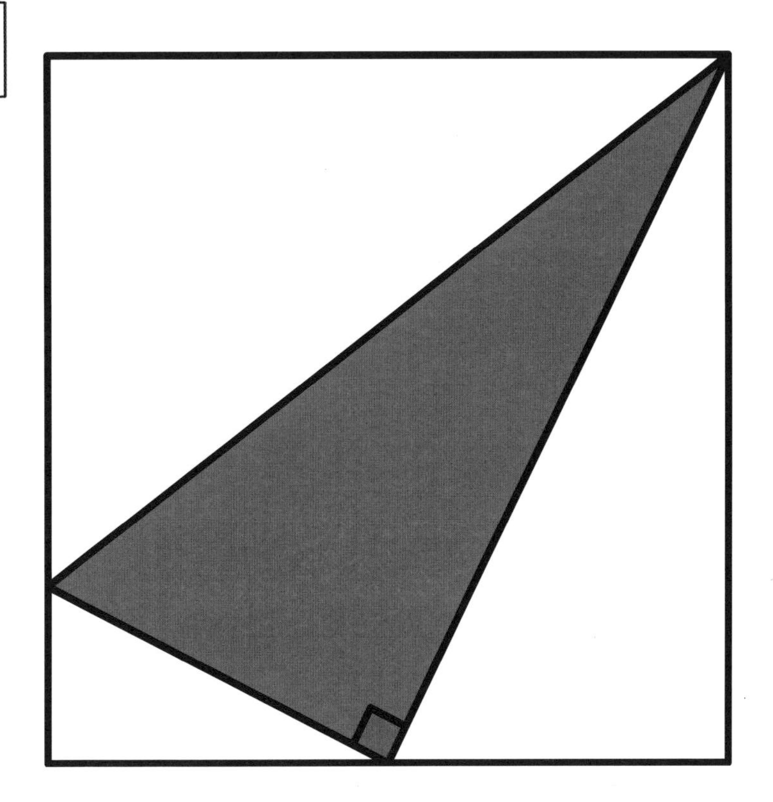

A right-angled triangle is constructed using the midpoint and vertex of a square as shown.

What fraction of the square is shaded?

Solutions and Answers

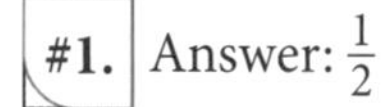

#1. Answer: $\frac{1}{2}$

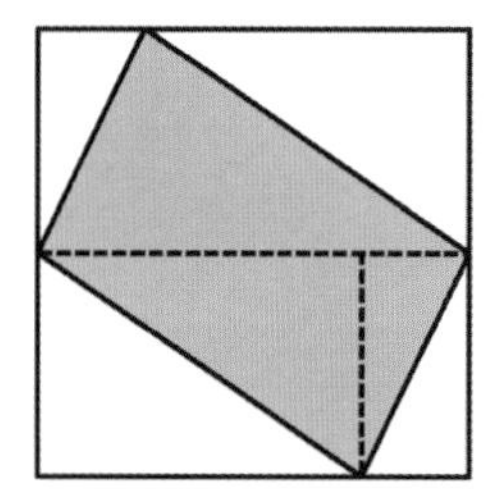

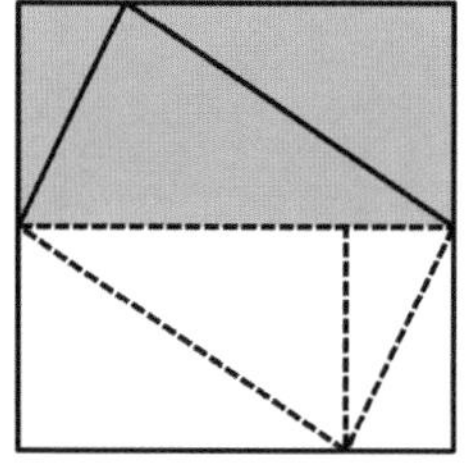

• Dissect the bottom part in two right triangles that fill the white space above.

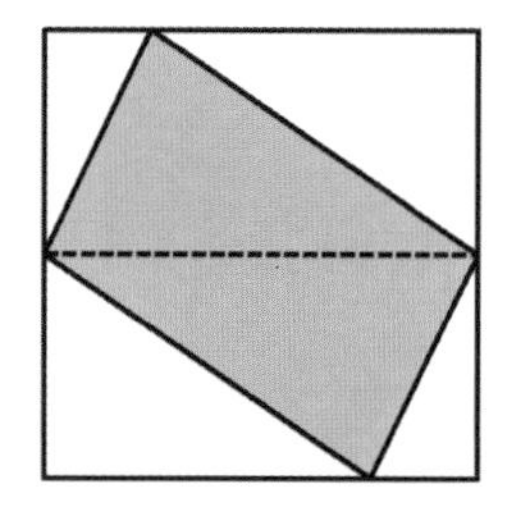

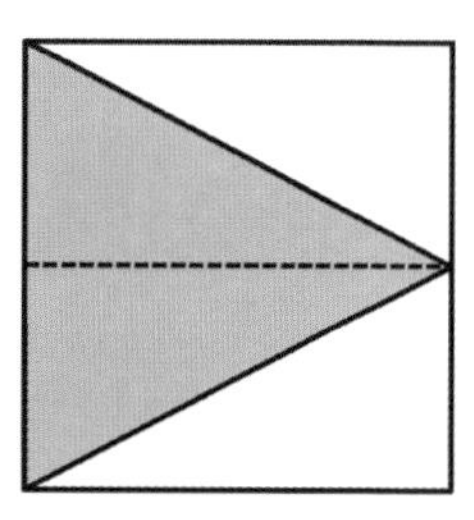

•• By shearing the shape, you can see that the area is equal to this triangle.

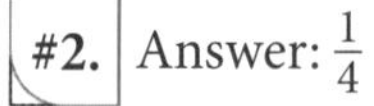

#2. Answer: $\frac{1}{4}$

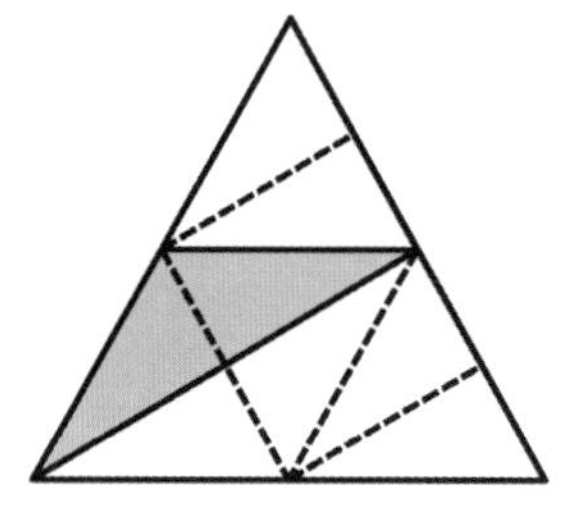

Dissection: $\frac{2}{8}$

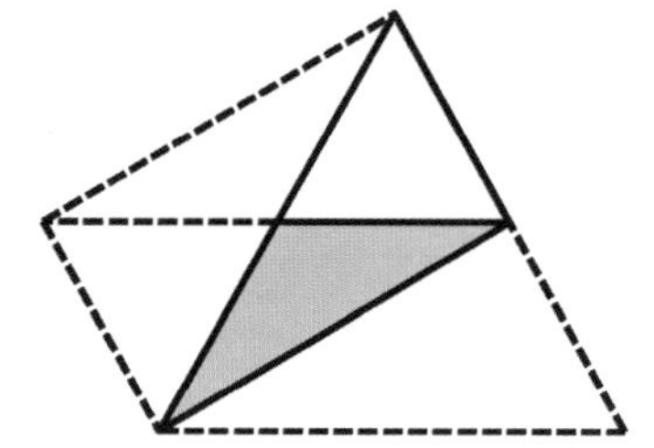

Seeing an envelope.

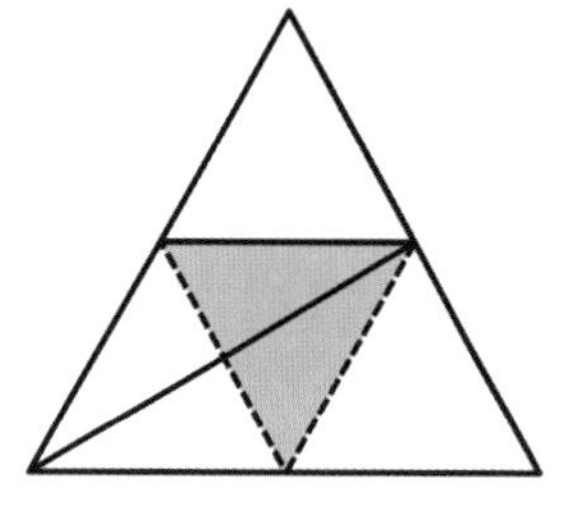

Same height and base.

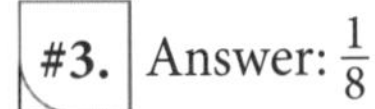

#3. Answer: $\frac{1}{8}$

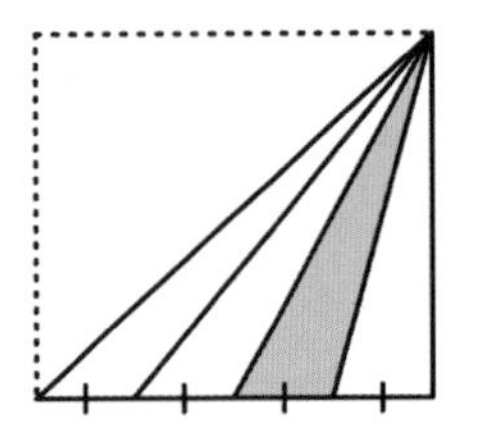

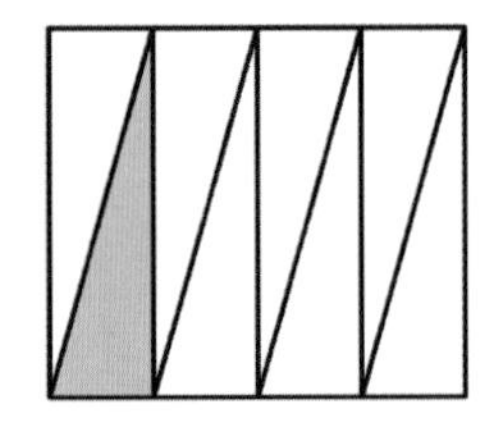

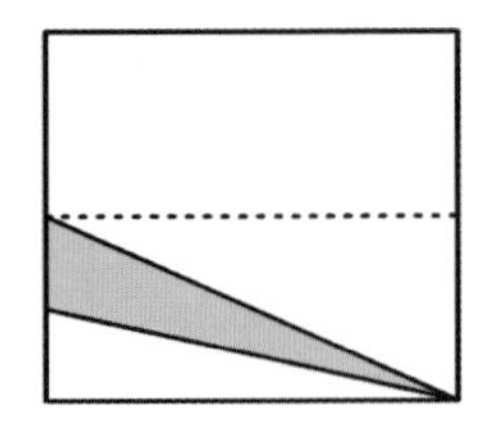

The eight triangles have the same base (a quarter of a side) and same altitude (side of the square). So they all have the same area. The shaded area represents one eighth of the square.

#4. Answer: $\frac{5}{8}$

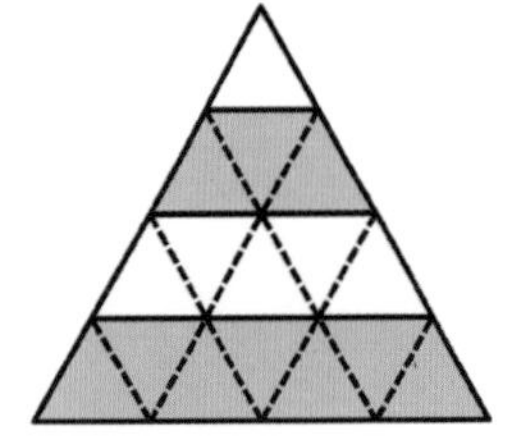

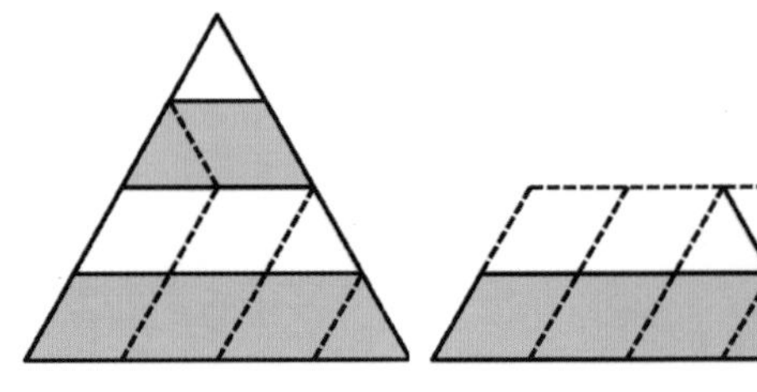

• Dissect one sixth of the figure into equilateral triangles: $\frac{10}{16}$.

•• Dissect into congruent rhombuses and rearrange to see $\frac{5}{8}$.

••• If you say the central hexagon has a unit area, the others have areas of 2^2, 3^2 and 4^2. You then calculate: $\frac{1}{16}(4^2 - 3^2 + 2^2 - 1) = \frac{10}{16} = \frac{5}{8}$

#5. Answer: $\frac{3}{4}$

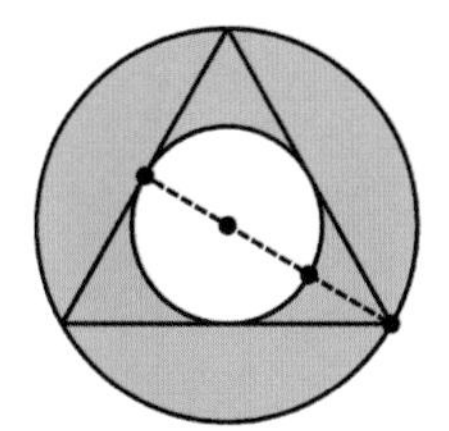

The centre of the circles is the centroid of the equilateral triangle.
The dots on the figure split the median in thirds.
The radii are in ratio 1:2, hence the shaded area has a ratio of:

$$\frac{2^2 - 1^1}{2^2} = \frac{3}{4}$$

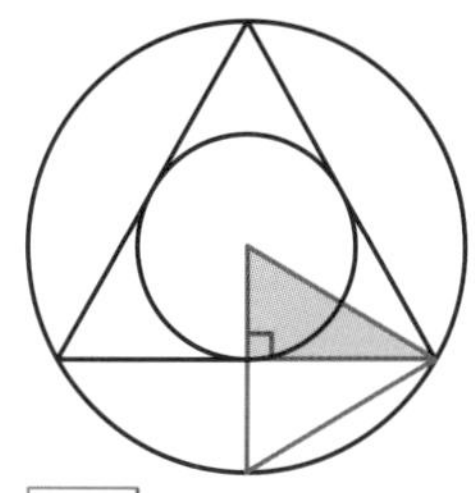

This figure highlights half an equilateral triangle from which you can deduce that the ratio of the radii is 1:2.

#6. Answer: $\frac{2}{5}$

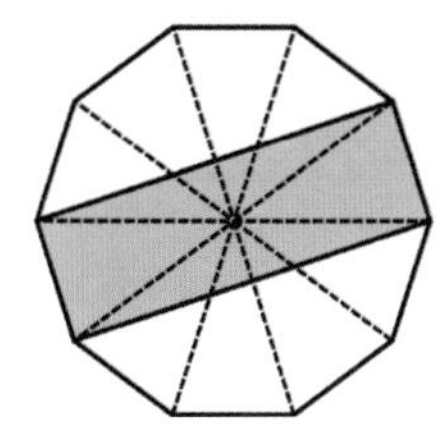

Dissect and rearrange the pieces of the rectangle into four of the ten congruent triangles.

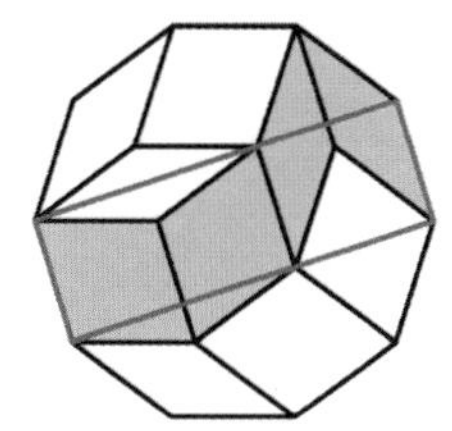

The rectangle has the same area as these four grey rhombuses. There are two of each kind in grey and five of each kind in the decagon.

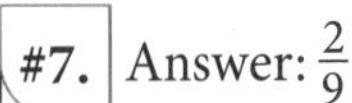

#7. Answer: $\frac{2}{9}$

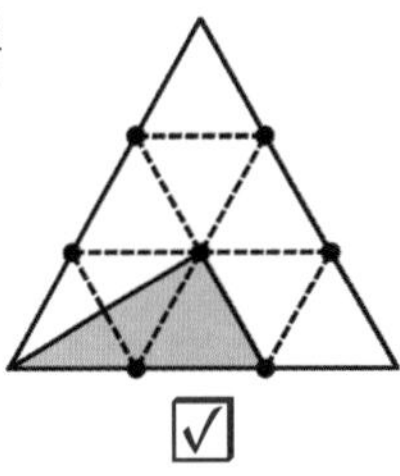

☑

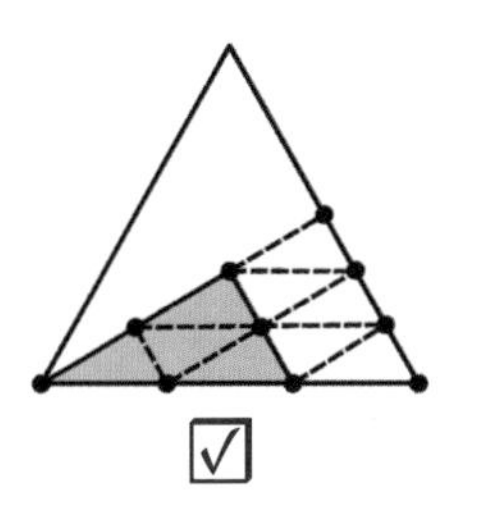

☑

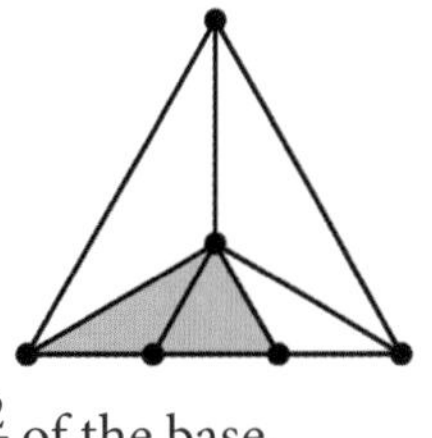

$\frac{2}{3}$ of the base,

$\frac{1}{3}$ of the height: $\frac{2}{3} \times \frac{1}{3} = \frac{2}{9}$

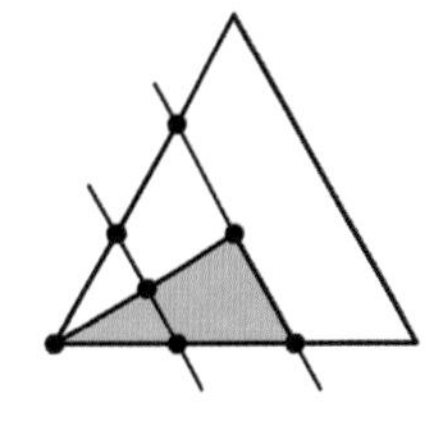

Half an equilateral triangle whose side is $\frac{2}{3}$ of the big one:

$$\frac{1}{2} \times \left(\frac{2}{3}\right)^2 = \frac{2}{9}$$

#8. Answer: $\frac{3}{4}$

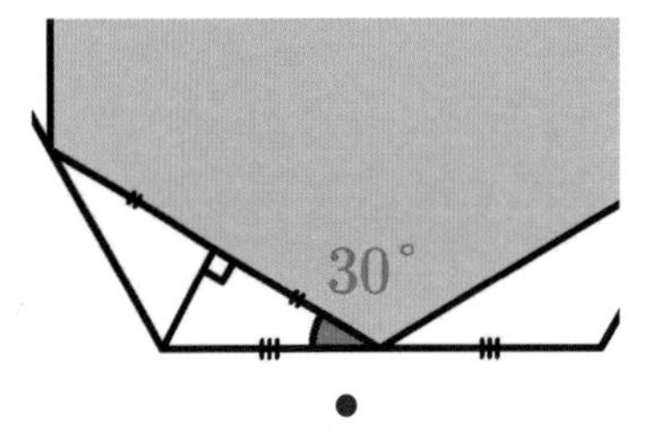

●

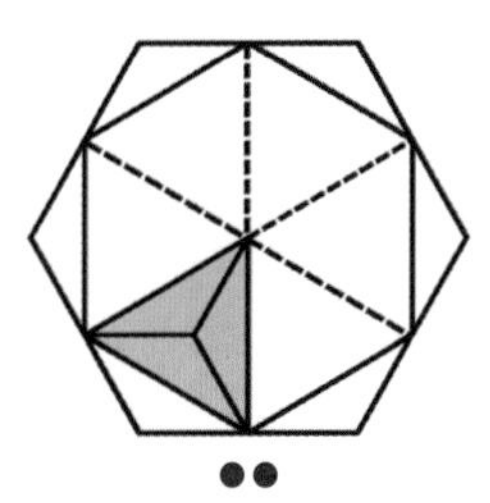

●●

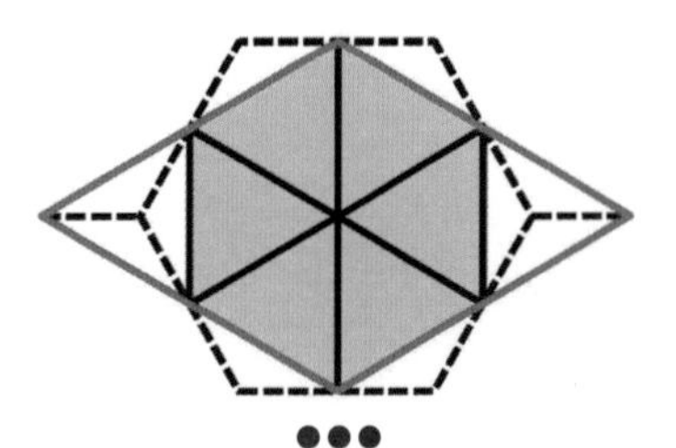

●●●

- ● The scaling factor to go from the large to the small hexagon is $\cos(30) = \frac{\sqrt{3}}{2}$ so the ratio of areas is its square: $\frac{3}{4}$.

- ●● The small hexagon is dissected into $6 \times 3 = 18$ grey isosceles triangles. Add six of those to form the large hexagon. $\frac{18}{24} = \frac{3}{4}$

- ●●● Moving the isosceles triangles you can count the equilateral triangles. $\frac{6}{8} = \frac{3}{4}$

#9. Answer: $\frac{1}{4}$

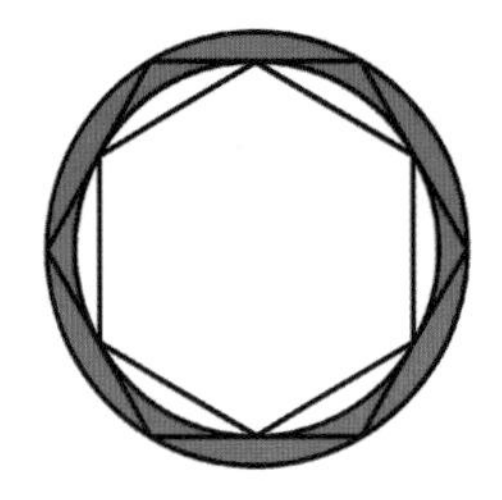

As seen in the previous snack, the scaling factor to go from the large to the small circle is $\cos(30) = \frac{\sqrt{3}}{2}$ so the ratio of areas of the two discs is its square: $\frac{3}{4}$, hence the shaded area is the remaining $\frac{1}{4}$.

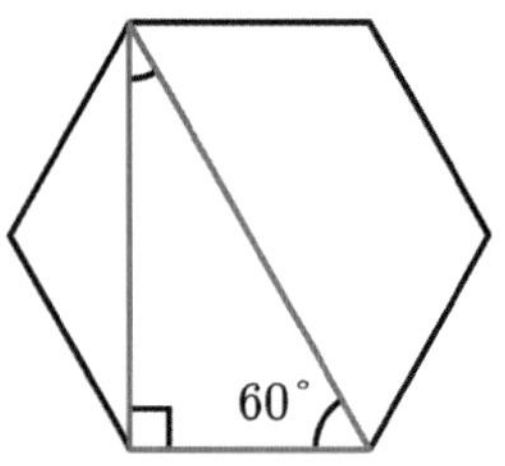

The altitude and hypotenuse of the triangle indicate the diameters of the incircle and circumcircle. The ratio of the sides of the triangle is $1 : \sqrt{3} : 2$ so their square's ratio is $1 : 3 : 4$. So the ratios of the areas of the circles is $3 : 4$.

#10. Answer: $\frac{1}{2}$

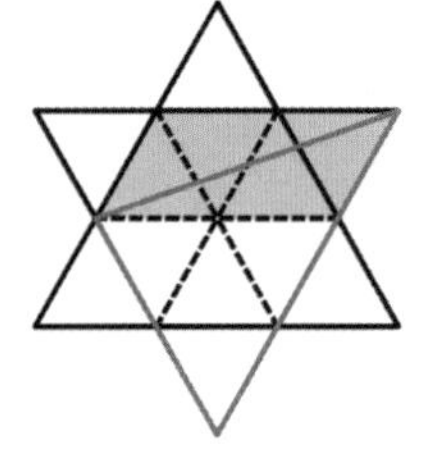

Dissect and count the equilateral triangles: $4 + \frac{1}{2} \times 4 = 6$.

$$\frac{6}{12} = \frac{1}{2}$$

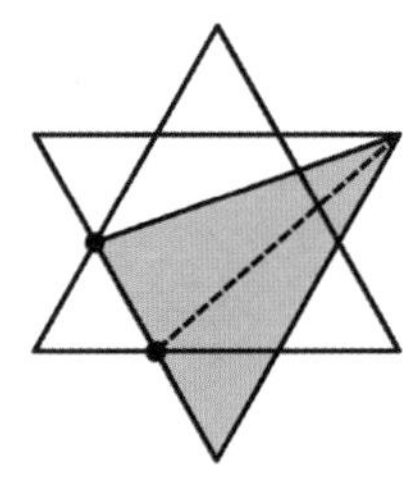

The base is $\frac{2}{3}$ of the equilateral triangle, with same height. So the grey area corresponds to $\frac{2}{3} = \frac{6}{9}$ of a big equilateral triangle.

That's 6 little equilateral triangles.

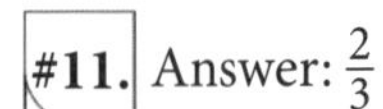

#11. Answer: $\frac{2}{3}$

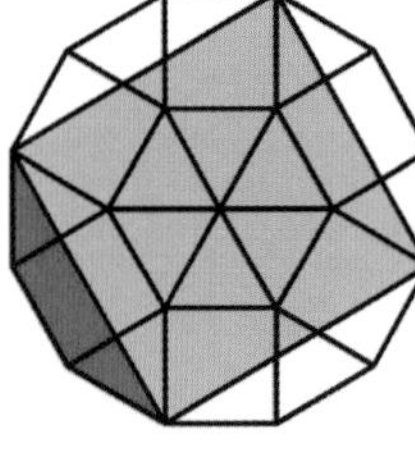

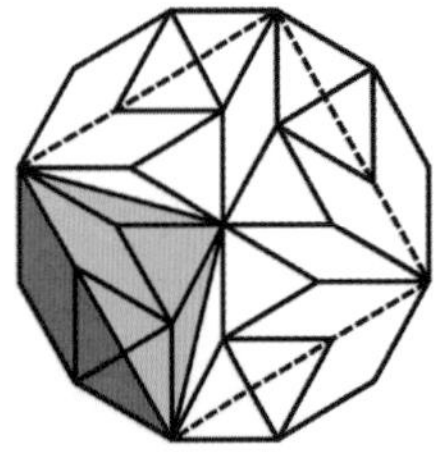

In this dissection, the dodecagon is made of 6 squares and 12 equilateral triangles ($6S + 12T$). Remove four times the red zone: $4\left(\frac{S}{2} + T\right)$ to get the grey square:

$$4S + 8T = \frac{2}{3} \times (6S + 12T).$$

Using the area R of a rhombus and T for an equilateral triangle, here a quarter of the figure shows $3R + 3T$ in the dodecagon and $2R + 2T$ in the square. That's $\frac{2}{3} \times (3R + 3T)$.

#12. Answer: $\frac{5}{16}$

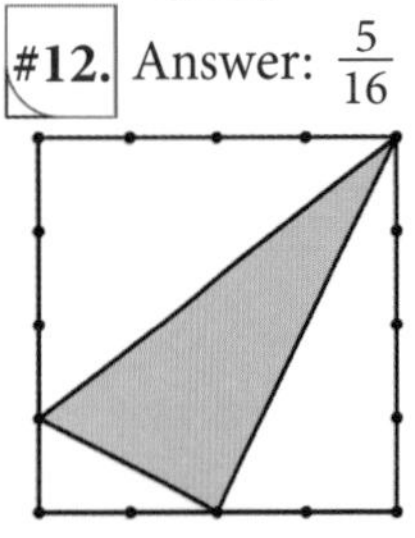

●

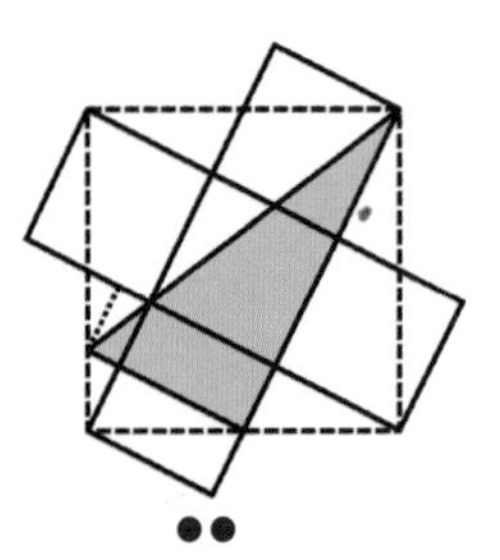

●●

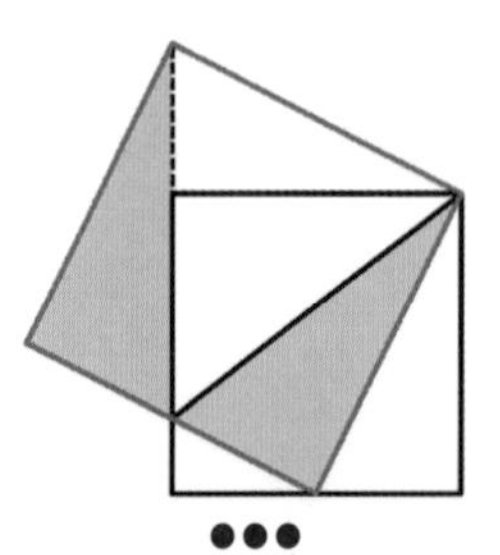

●●●

● Say the square has a side of 4 units. The two bottom white right triangles are similar (check the complementary angles) so by subtracting the three white right triangles you get: $1 - \frac{1}{2}(4 \times 3 + 4 \times 2 + 2 \times 1)/16 = \frac{5}{16}$

●● The area of the square is the same as that of the cross made of five little squares that we choose as units. The grey area is made of half of two squares, half a square and a quarter of a quarter of a square. The ratio is: $\left(1 + \frac{1}{2} + \frac{1}{16}\right)/5 = \frac{5}{16}$

●●● If the initial square has unit side, the new one which is half shaded has an area of $1^2 + \left(\frac{1}{2}\right)^2 = \frac{5}{4}$. Therefore, the fraction is: $\frac{1}{4} \times \frac{5}{4} = \frac{5}{16}$.

What's the Angle?

The aim of these puzzles is to find the value of the highlighted angle in each figure. Any assumptions required to solve the puzzles are indicated in the text below each figure.

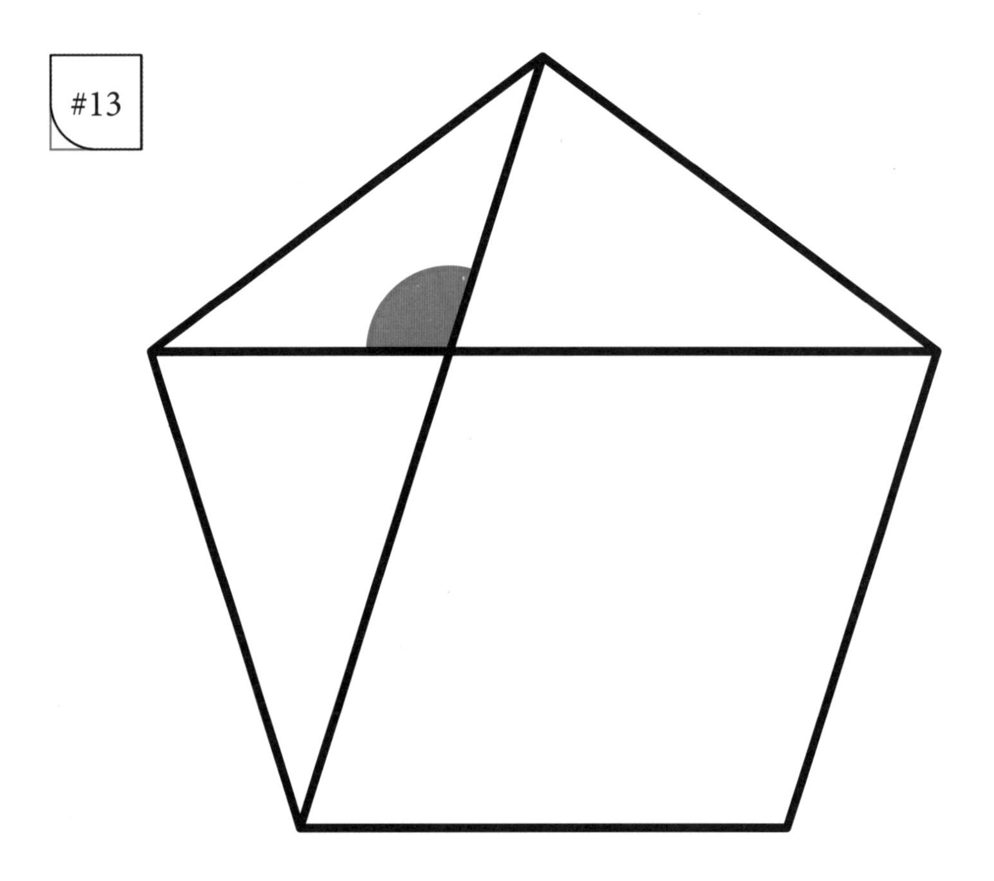

This is a regular pentagon. Find the value of the missing angle.

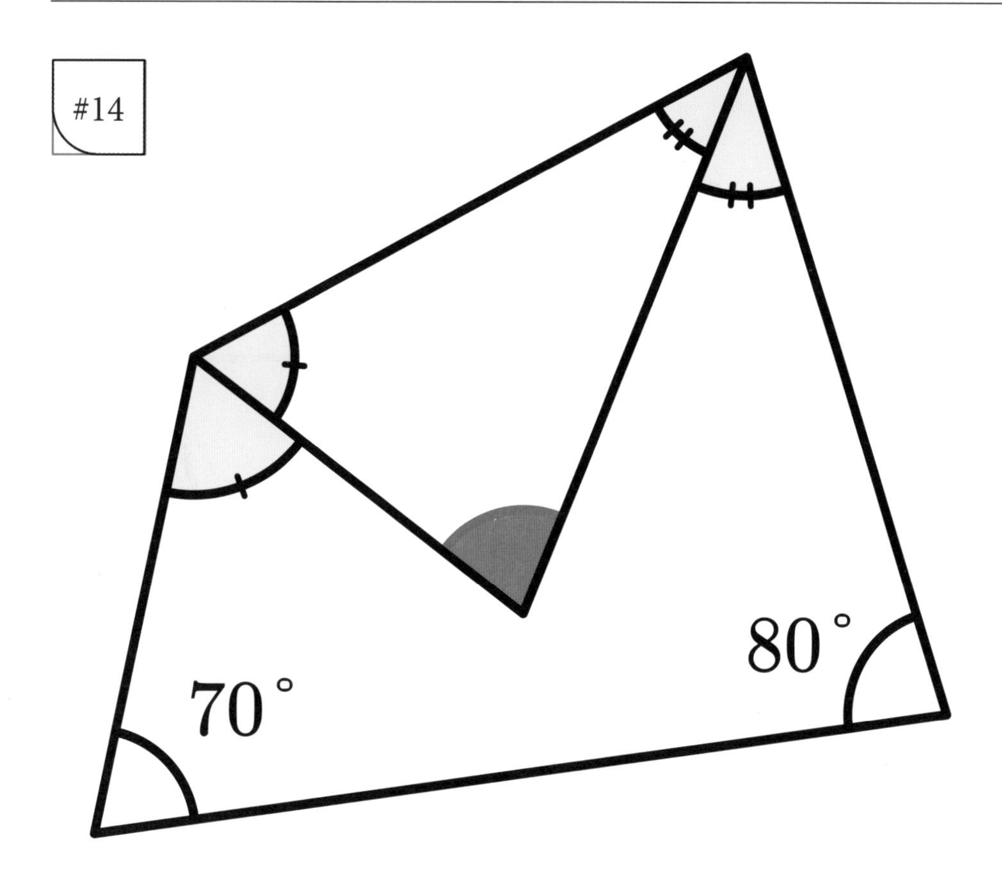

An irregular quadrilateral is constructed as shown.

Find the value of the missing angle.

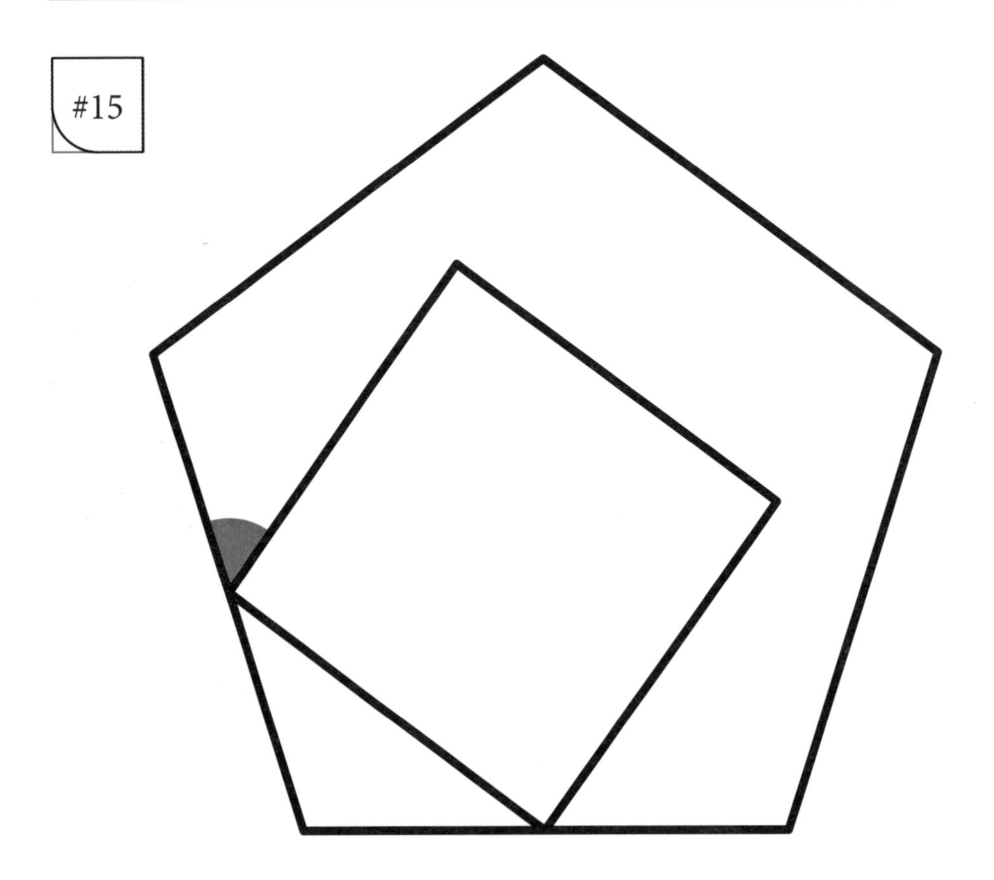

A square is constructed using the midpoints of two sides of a regular pentagon as shown.

Find the value of the missing angle.

#16

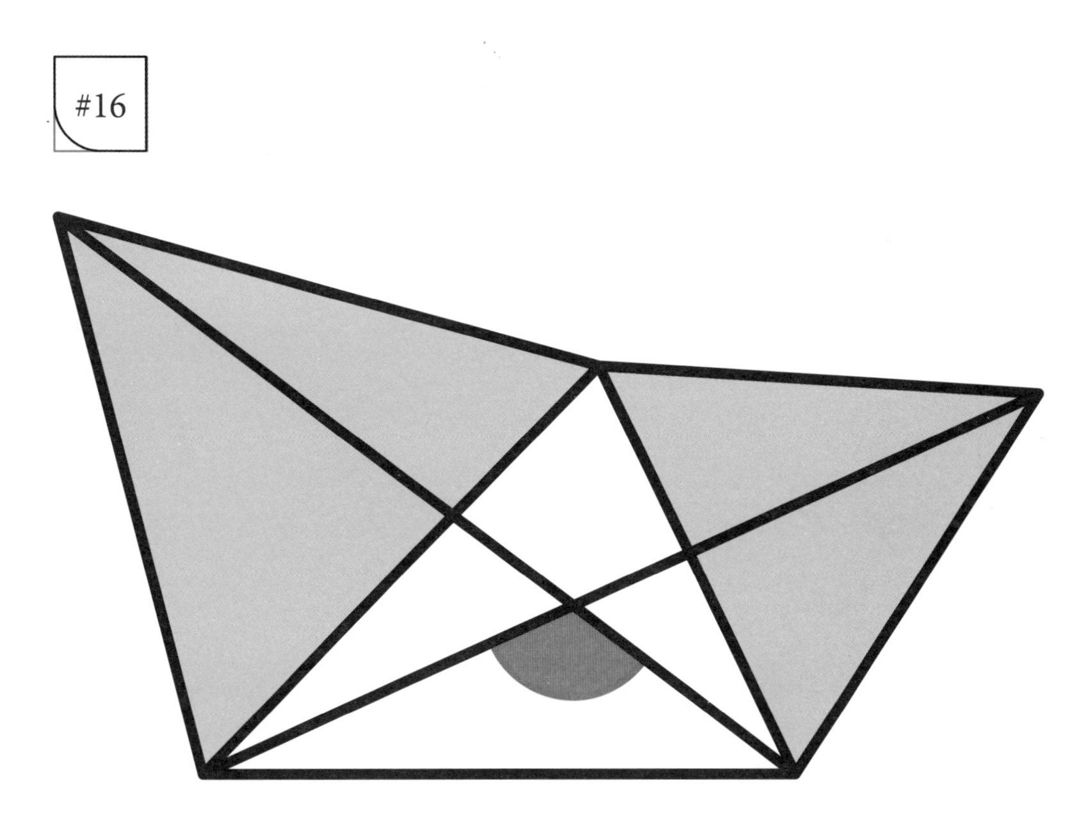

Two equilateral triangles are constructed such that they share a side with a third triangle as shown.

Find the value of the missing angle.

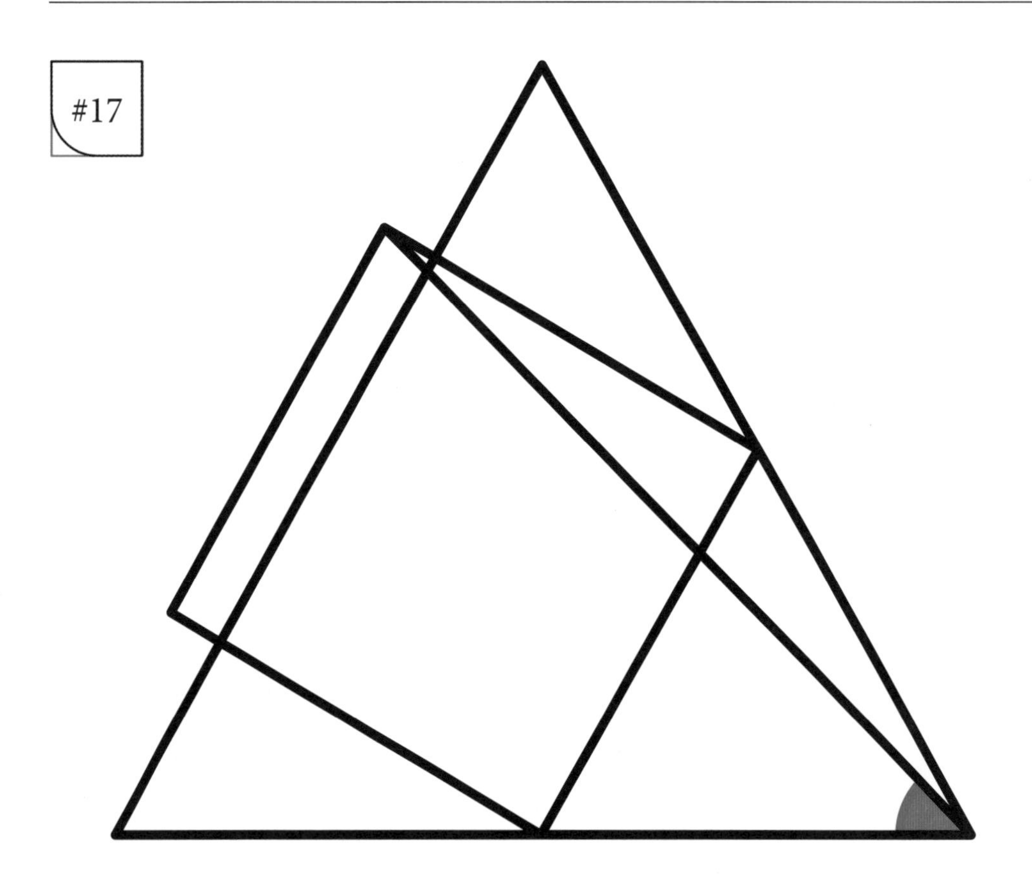

A square is constructed using the midpoints of two sides of an equilateral triangle as shown.

Find the value of the missing angle

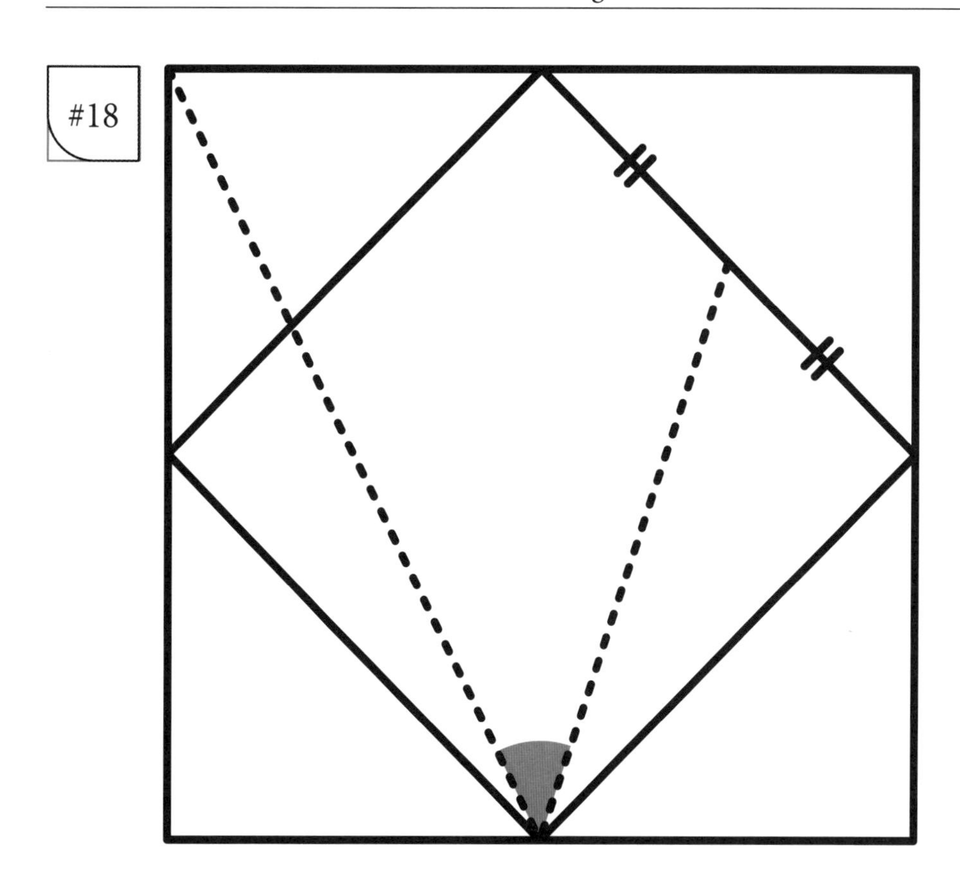

A square is constructed using the midpoints of a second square as shown.

Find the value of the missing angle.

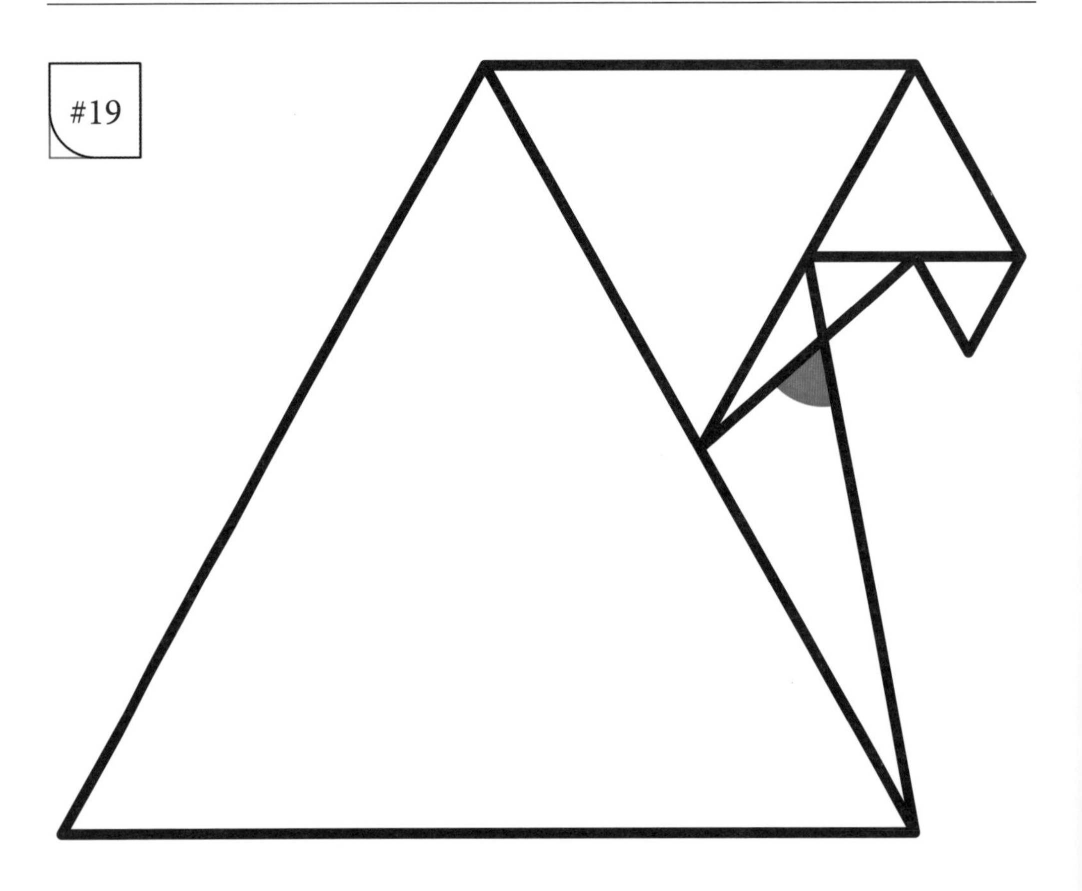

A sequence of equilateral triangles is constructed using vertices and midpoints as shown.

Find the value of the missing angle.

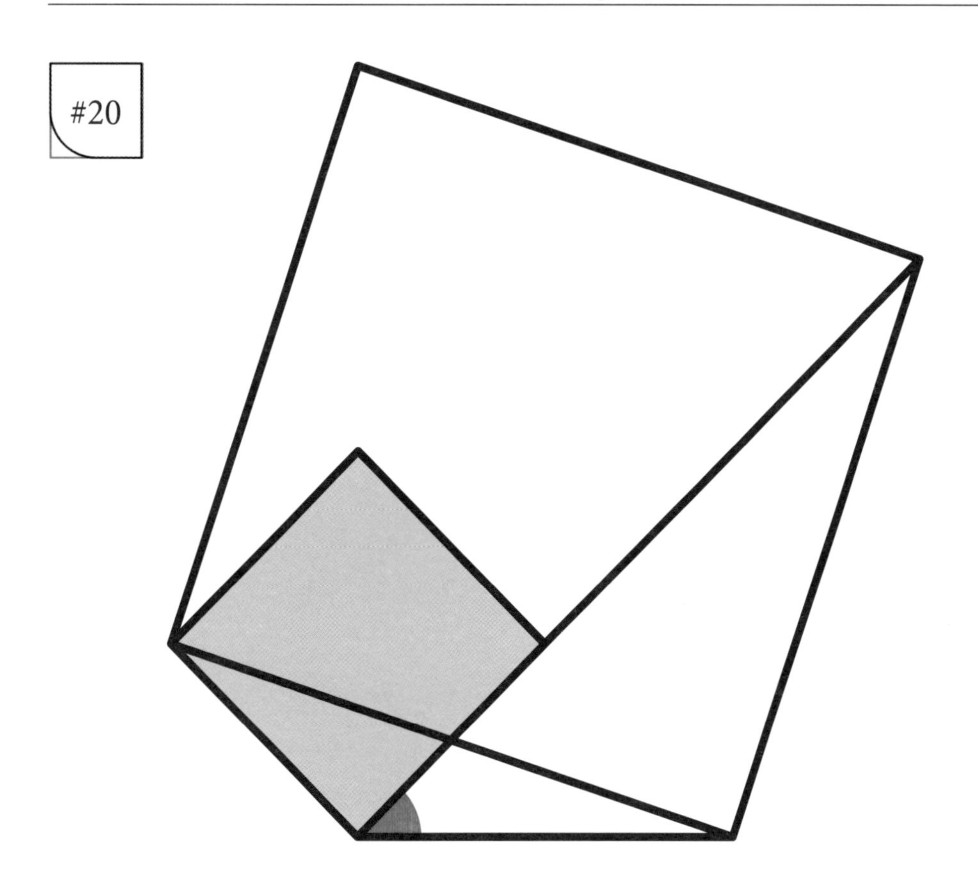

Two squares are constructed such that three vertices are collinear as shown.

Find the value of the missing angle.

#21

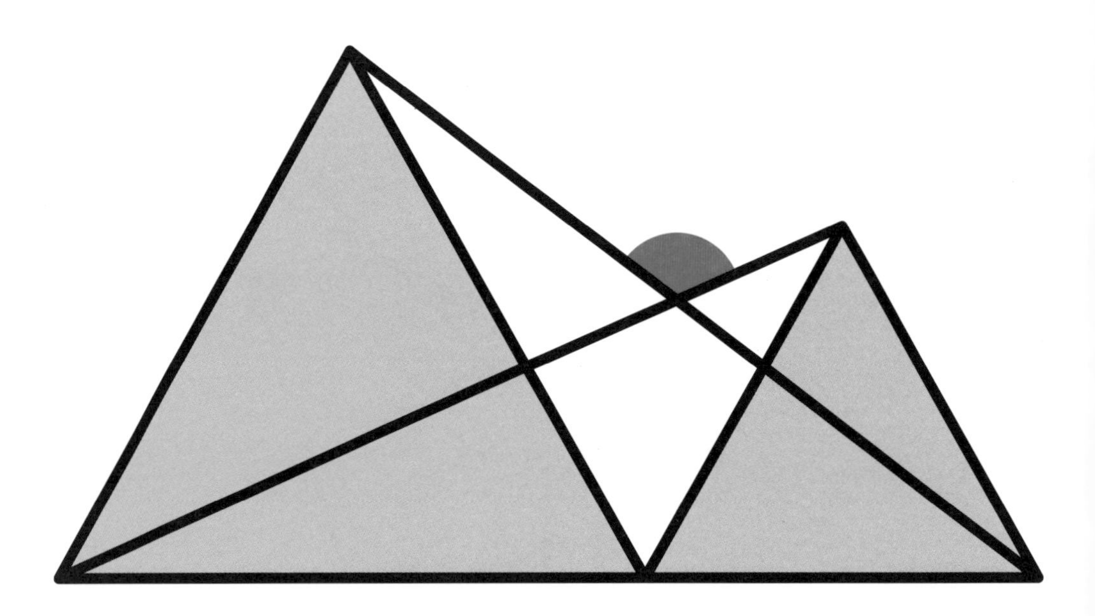

Two equilateral triangles touch at a vertex as shown.

Find the value of the missing angle.

#22

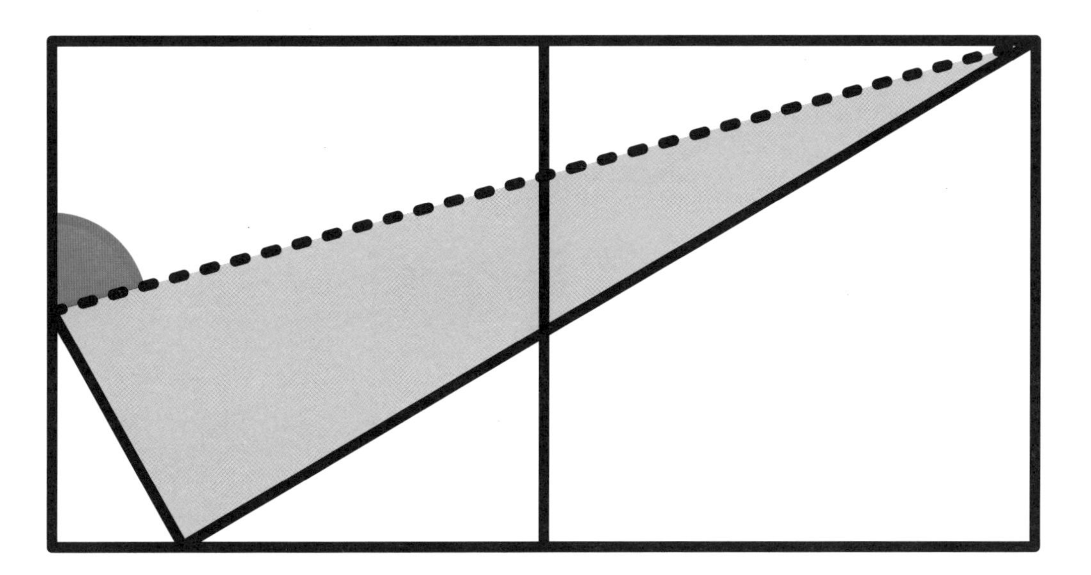

A rectangular sheet of paper made of two squares is folded such that one vertex touches the opposite side as shown.

Find the value of the missing angle.

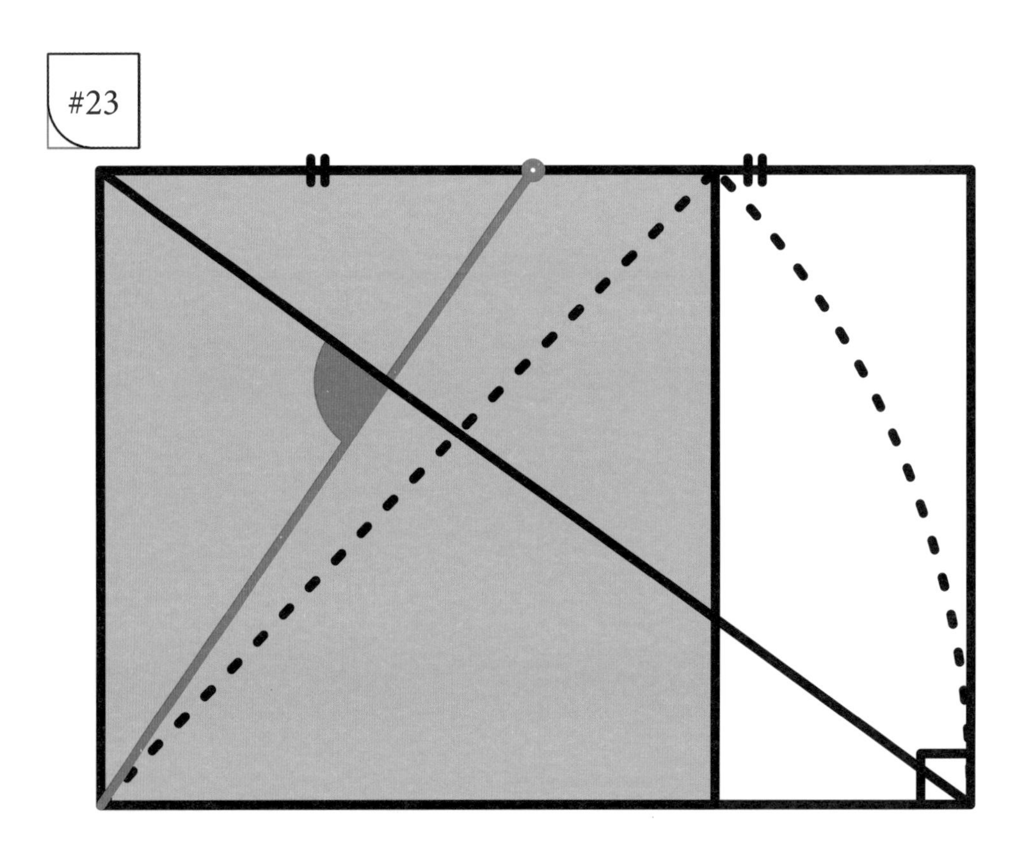

A square and a rectangle are constructed using a circle sector as shown.

Find the value of the missing angle.

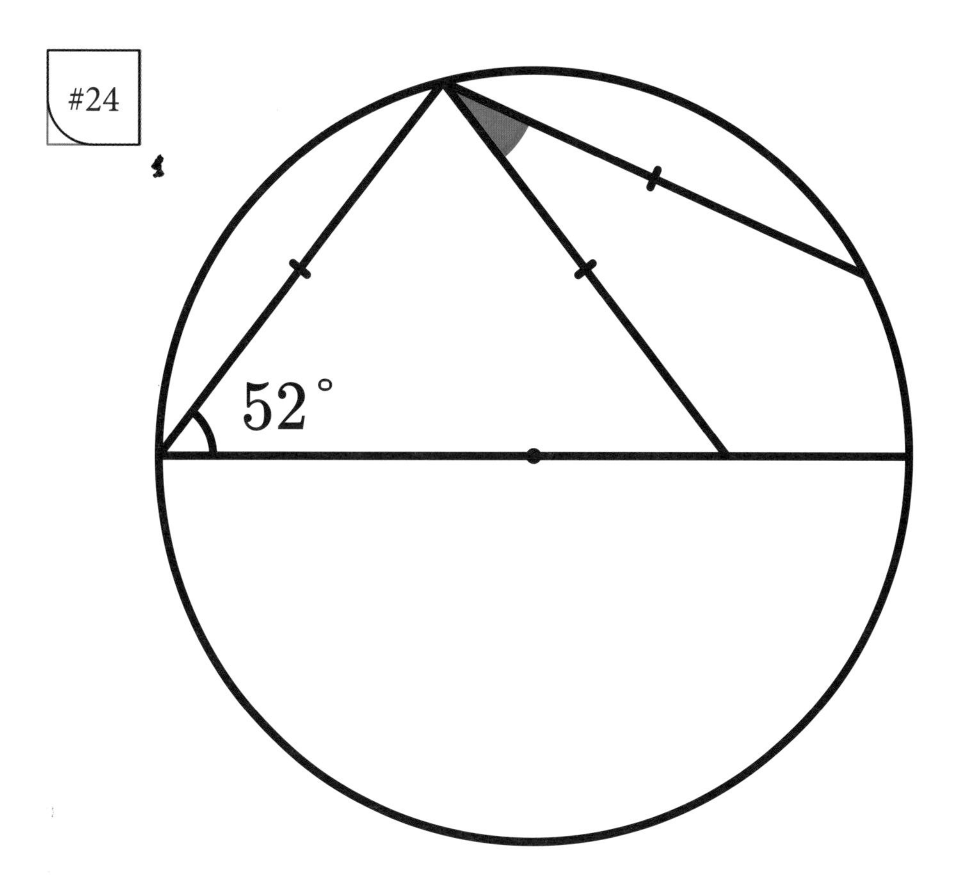

Three equal line segments are constructed from the same point on a circle as shown.

Find the value of the missing angle.

Solutions and Answers

#13. Answer: 108°

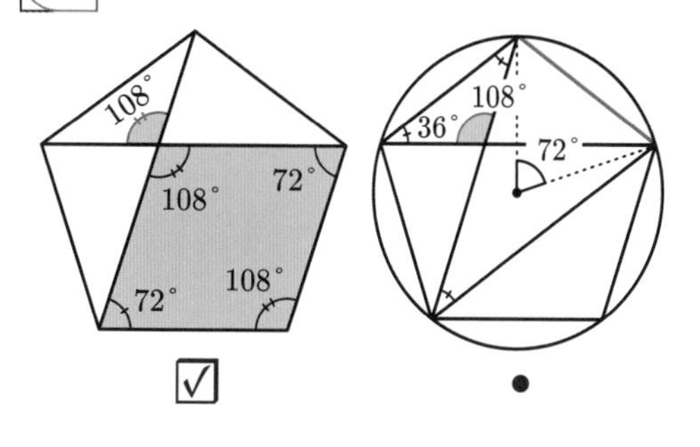

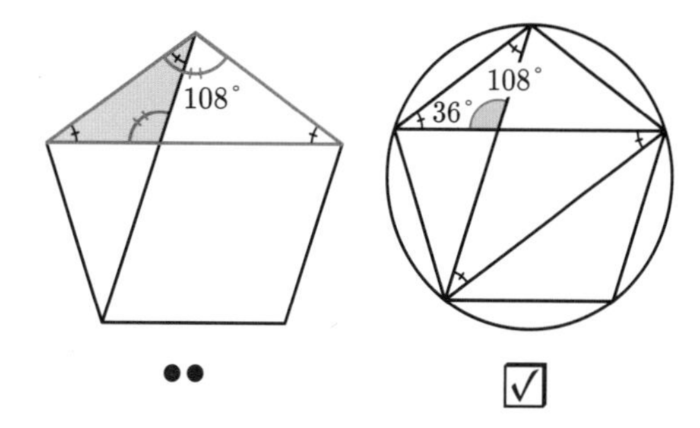

● By circumscribing the pentagon, we can deduce that the angle at the centre is double the angle at the circumference.

●● The isosceles triangles indicated are similar.

#14. Answer: 75°

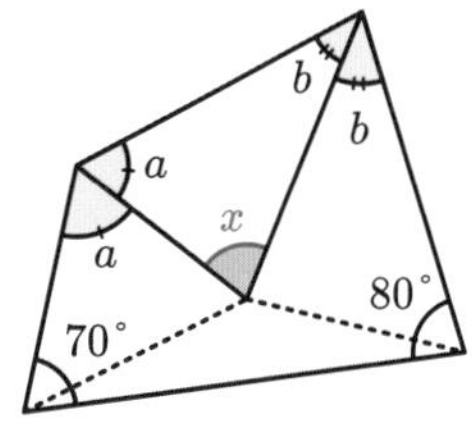

In the triangle on the top: $a + b + x = 180$ (1)
Adding the angles in the three bottom triangles:
$a + 70 + 80 + b + 360 - x = 3 \times 180 \Leftrightarrow a + b - x = 30$ (2)
(1) – (2) : $2x = 150$ so $x = 75°$

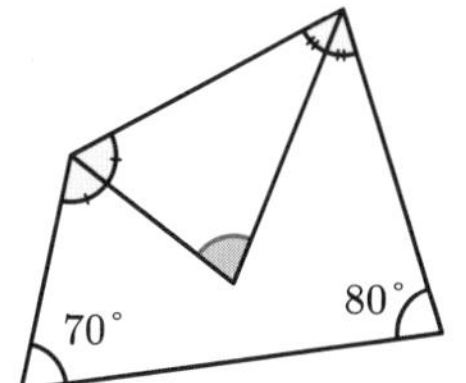

Each missing angle of the quadrilateral is constructed as two angles which we will call $a + b$.
$2a + 2b = 360 - 150 = 210$ so $a + b = 105$ and so $x = 75°$

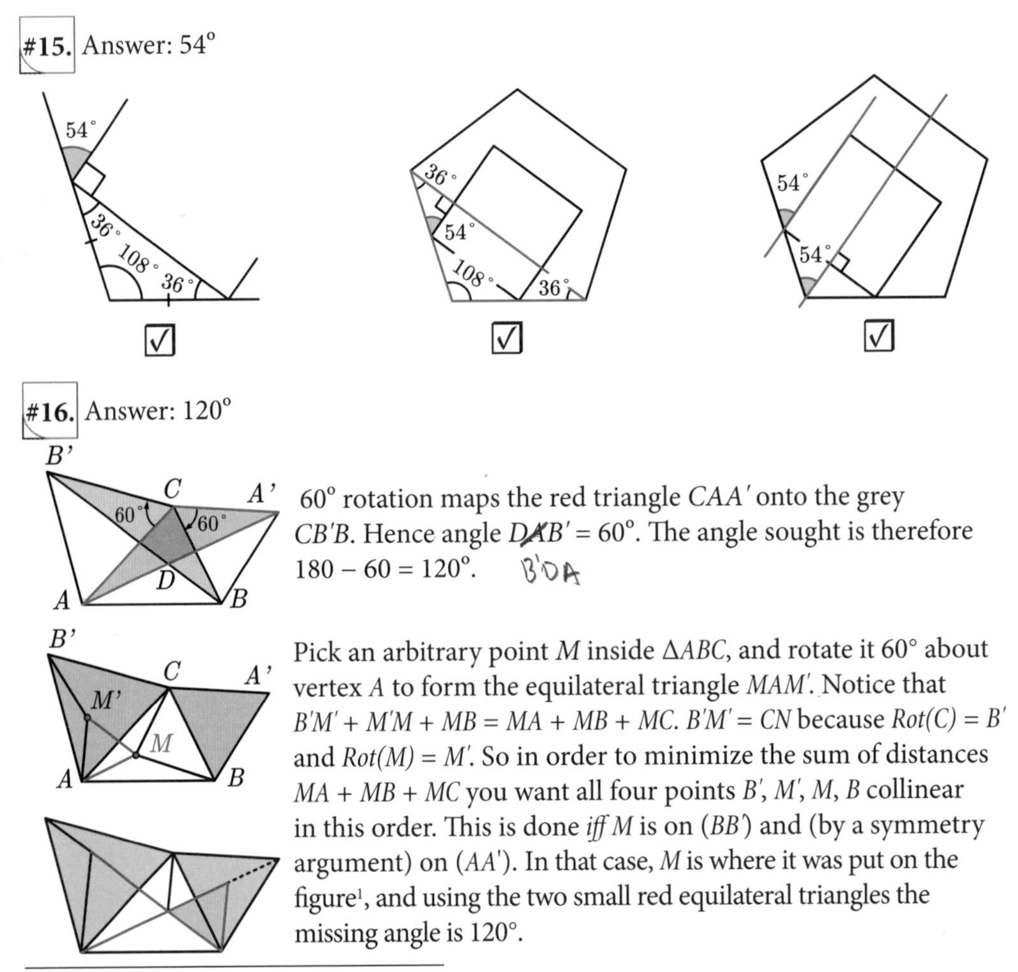

#15. Answer: 54°

#16. Answer: 120°

60° rotation maps the red triangle CAA' onto the grey $CB'B$. Hence angle $DAB' = 60°$. The angle sought is therefore $180 - 60 = 120°$.

Pick an arbitrary point M inside ΔABC, and rotate it 60° about vertex A to form the equilateral triangle MAM'. Notice that $B'M' + M'M + MB = MA + MB + MC$. $B'M' = CN$ because $Rot(C) = B'$ and $Rot(M) = M'$. So in order to minimize the sum of distances $MA + MB + MC$ you want all four points B', M', M, B collinear in this order. This is done *iff* M is on (BB') and (by a symmetry argument) on (AA'). In that case, M is where it was put on the figure[1], and using the two small red equilateral triangles the missing angle is 120°.

[1] This point which minimizes the sum of distances to the vertices of the triangle is known as the FERMAT-TORICELLI point. We just proved that from this point you see the vertices of the triangle with an angle of 120°.

#17. Answer: 45°

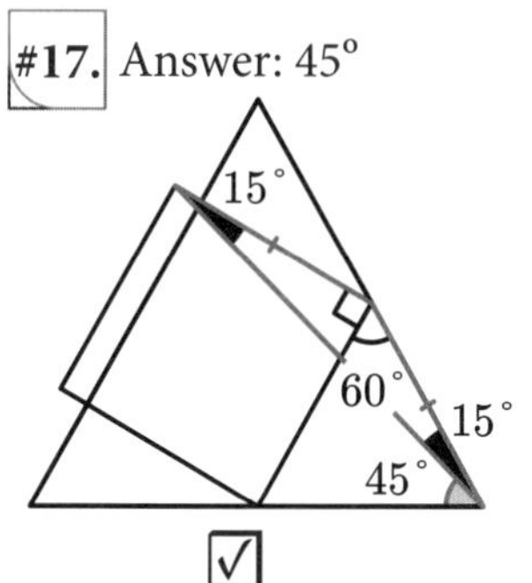

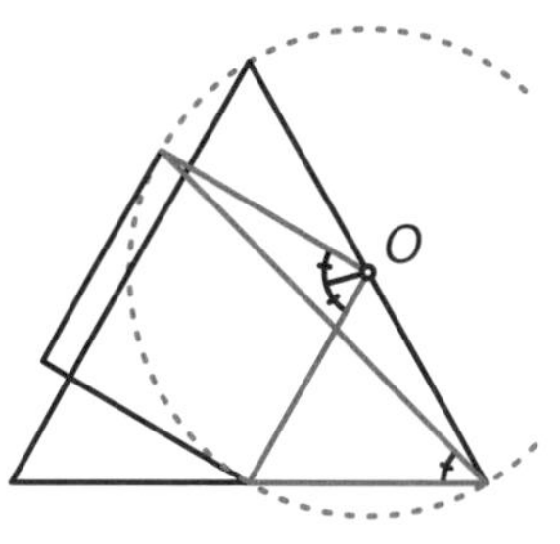

The point marked O is equidistant from the four other marked points that must therefore lie on a circle as shown. Using the central angle theorem gives $x = 90 \div 2 = 45°$.

#18. Answer: 45°

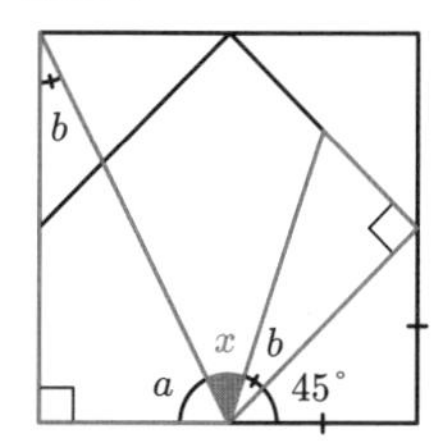

The two red triangles with right angles shown are similar (leg ratio 1:2) with $a + b = 90°$ so $a + x + b + 45 = 180°$ so $x = 45°$.

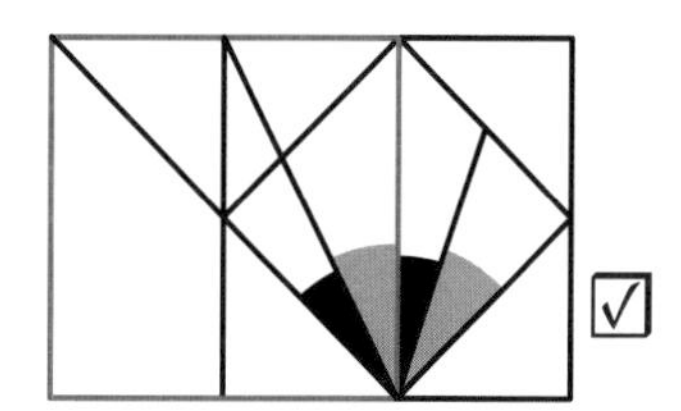

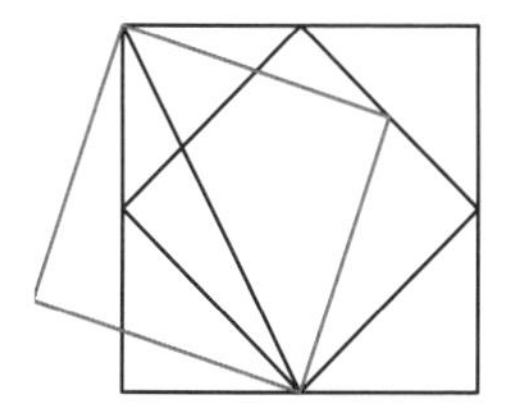

A combined rotation of 45° and scaling transforms the small black square onto the tilted square on the left.

#19. Answer: 120°

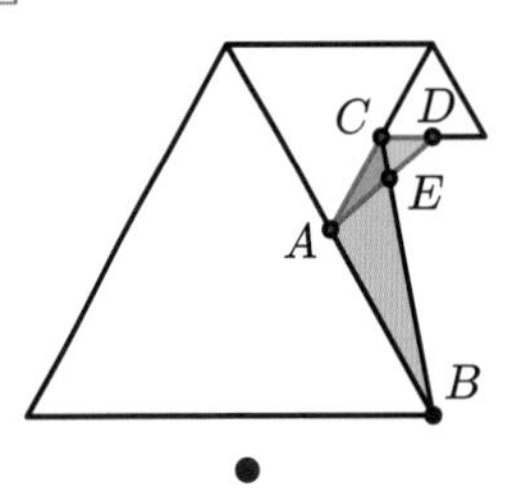

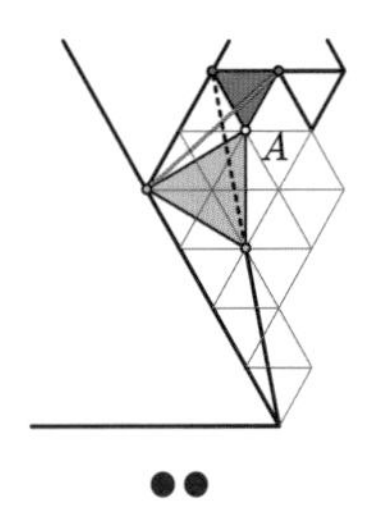

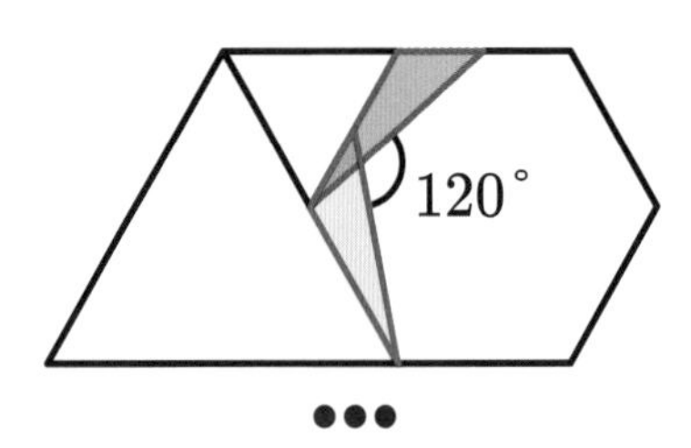

- ΔABC and ΔACD are similar: $\angle BAC = 120°$ and $\angle ACB + \angle DAC = 60°$, so $\angle AEC = 120°$, so $\angle AEB = 60°$.

●● Using the red grid you can spot the same pattern as in snack #16, where we saw that, a 60° rotation maps one line onto the other.

●●● Extending the lines we form two congruent triangles rotated by 120° as can be seen using the regular hexagon drawn.

#20. Answer: 45°

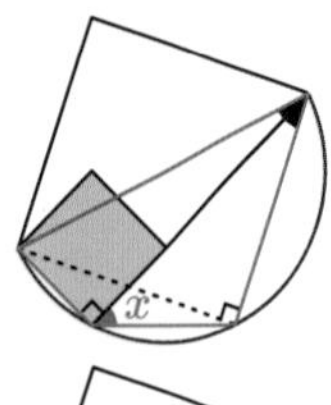

The right angles explain the semicircle. The black angle is 45°, and by looking at the red chord formed by the largest square, the angle which intercepts the same arc as the black angle but on the other side of the chord is $x + 90°$, so $x = 180 - 45 - 90 = 45°$

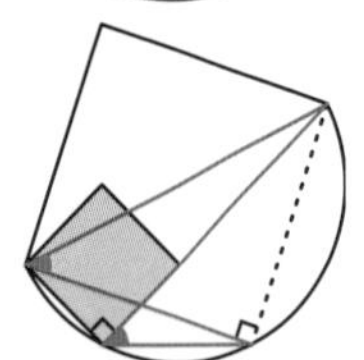

Same circle but using the "bow tie" shape you have the two congruent angles in red.

#21. Answer: 120°

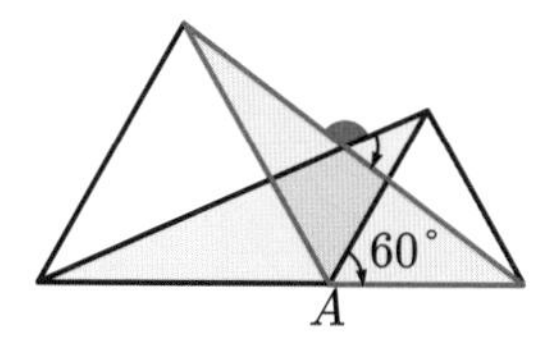

The red triangle on the right is a 60° rotation of the grey triangle on the left, centre A.

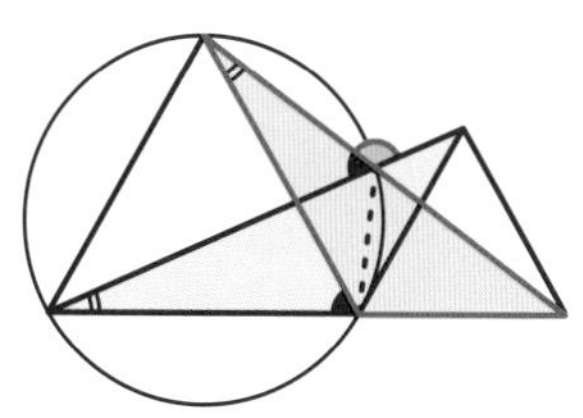

Congruent triangles form congruent angles (as indicated by the two dashes) so we have four cocyclic points and therefore the other angles of the 'bow tie' are also congruent.

An alternate approach could be to view the figure as a variant of the figure in Snack #16 with $\angle ACB = 180°$.

#22. Answer: 75°

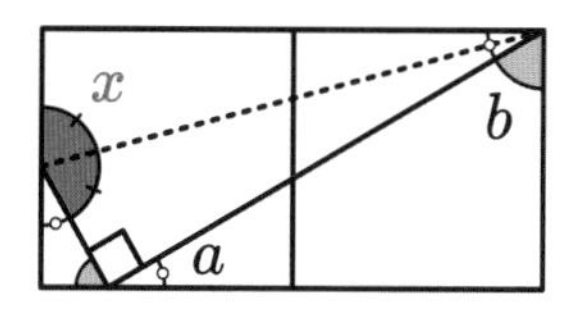

$\sin(a) = \frac{1}{2}$ so $a = 30°$.
Notice two similar right triangles and that $2x + a = 180°$ gives $x = 75°$

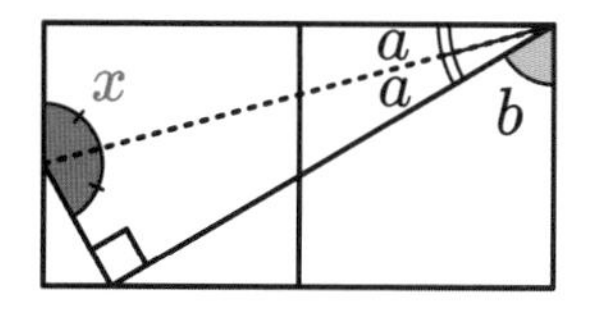

$\cos(b) = \frac{1}{2}$ so $b = 60°$ and $b + 2a = 90°$ gives $a = 15°$ so $x = 90 - 15 = 75°$

#23. Answer: 90°

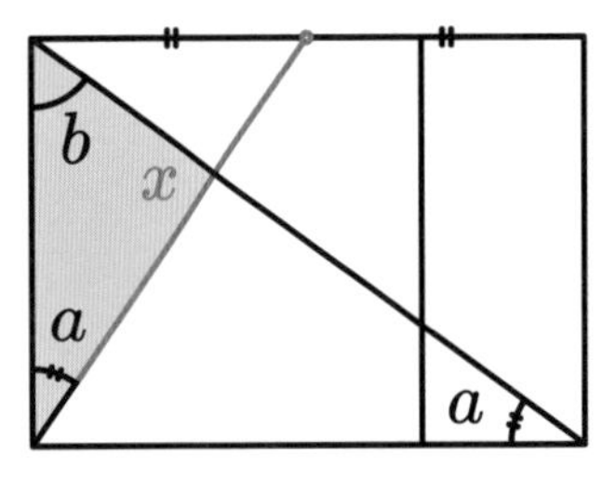

The two angles marked a are congruent because their tangent is the same: $\frac{1}{\sqrt{2}} = \frac{\sqrt{2}/2}{1}$. It follows that the red triangle with angles a, b, x is similar to the right triangle with acute angles a and b so $x = 90°$.

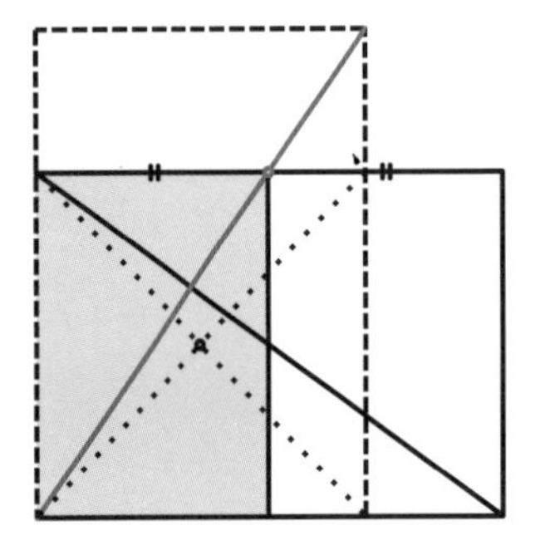

Give a quarter turn to the rectangle around the centre of the square to get the dashed rectangle. It is similar to the grey rectangle (because the ratios width over length are equal : $\frac{1}{\sqrt{2}} = \frac{\sqrt{2}/2}{1}$ so the red diagonal from the bottom left vertex is also that of the grey rectangle and is therefore perpendicular to the black diagonal.

#24. Answer: 28°

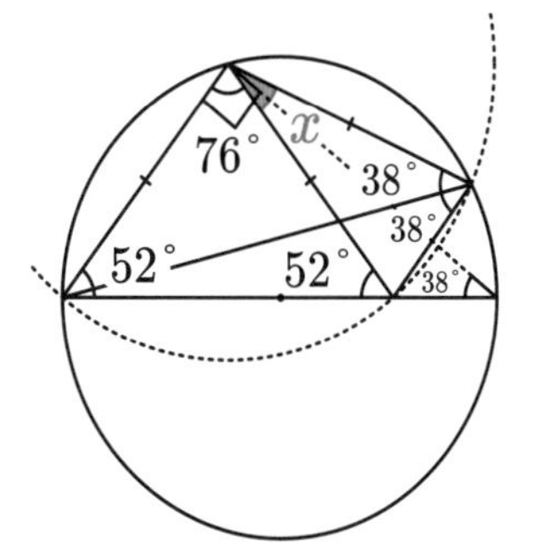

Looking at the dashed circle, x is a central angle and the right dashed triangles gives

$a + \frac{x}{2} = 90°$ (1).

From the semicircle angle a and $x + 76$ are on the both sides of the same arc so $x + 76 = 180 - a$ which gives

$a + x = 104°$ (2).

(2) − (1) gives $\frac{x}{2} = 14°$ hence $x = 28°$.

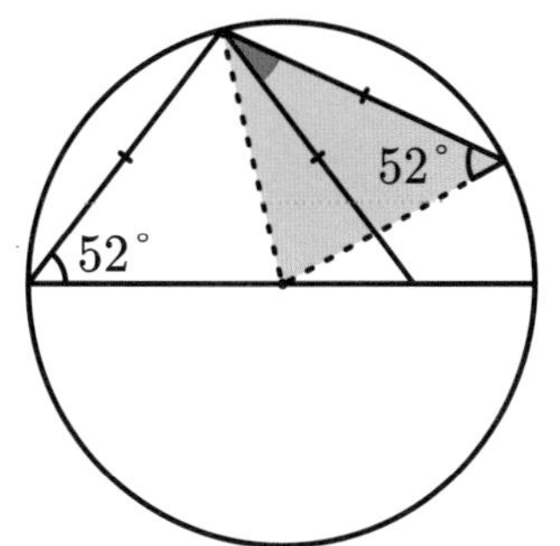

Notice three isosceles triangles with angles 52°, 52°, 76°. Looking at the angles at the point on the top:

$76 + x = 52 + 52$ so $x = 28°$

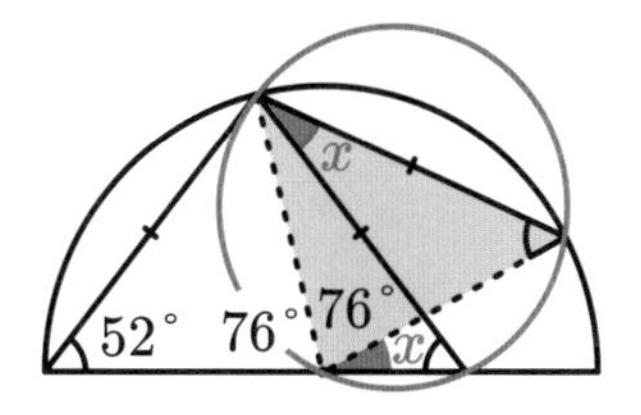

The grey triangle and the white triangle are congruent and isosceles with the same base (angles 52°, 52°, 76°). Marked lengths give two other isosceles triangles, which explains the three 52° angles marked. This justifies the existence of the red circle. So the angles marked x are congruent.

$x + 76 + 76 = 180° \Rightarrow x = 28°$.

Prove it!

The aim of these puzzles is to prove mathematically that the statements provided are true. Any assumptions required to solve the puzzles are indicated in the text below each figure.

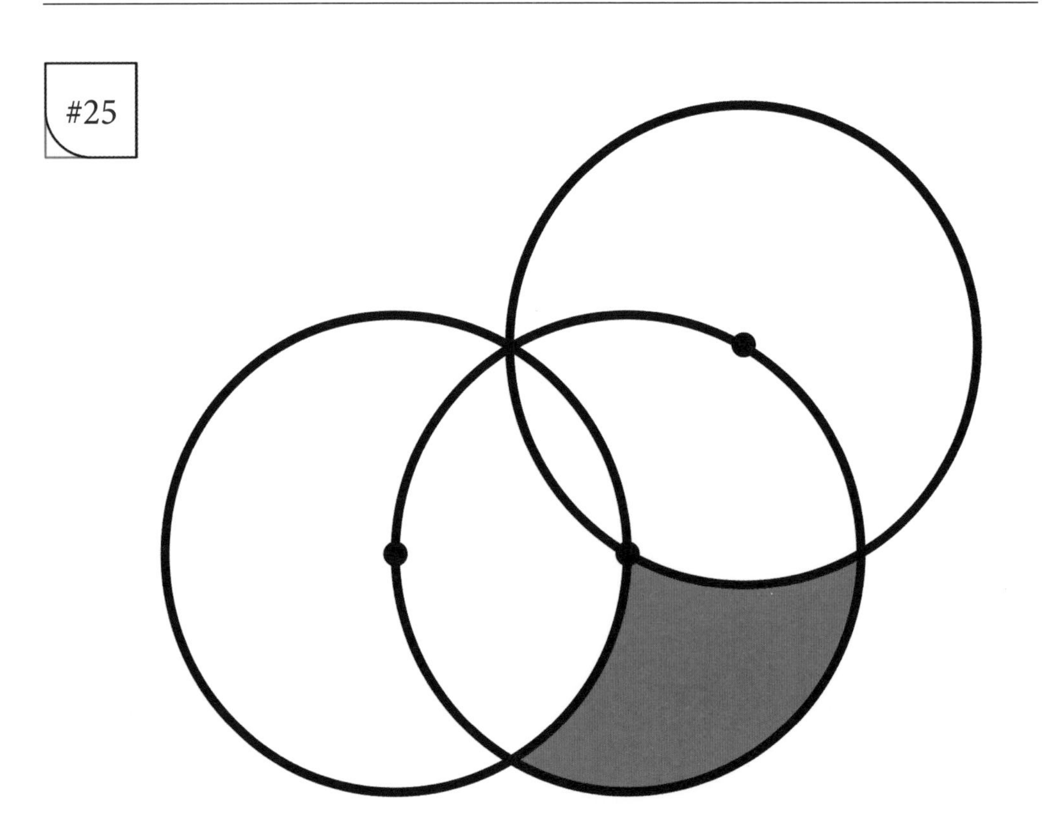

Three congruent circles are constructed as shown. Prove that finding the area of the shaded region does not require π.

#26

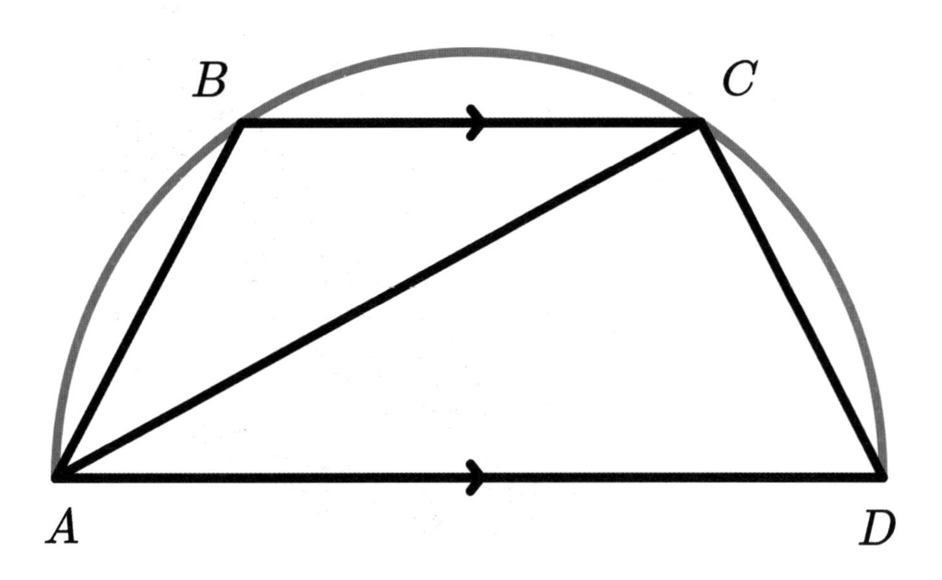

Using a semicircle as shown, prove that $\angle ABC - \angle BCA = \angle ACD$.

#27

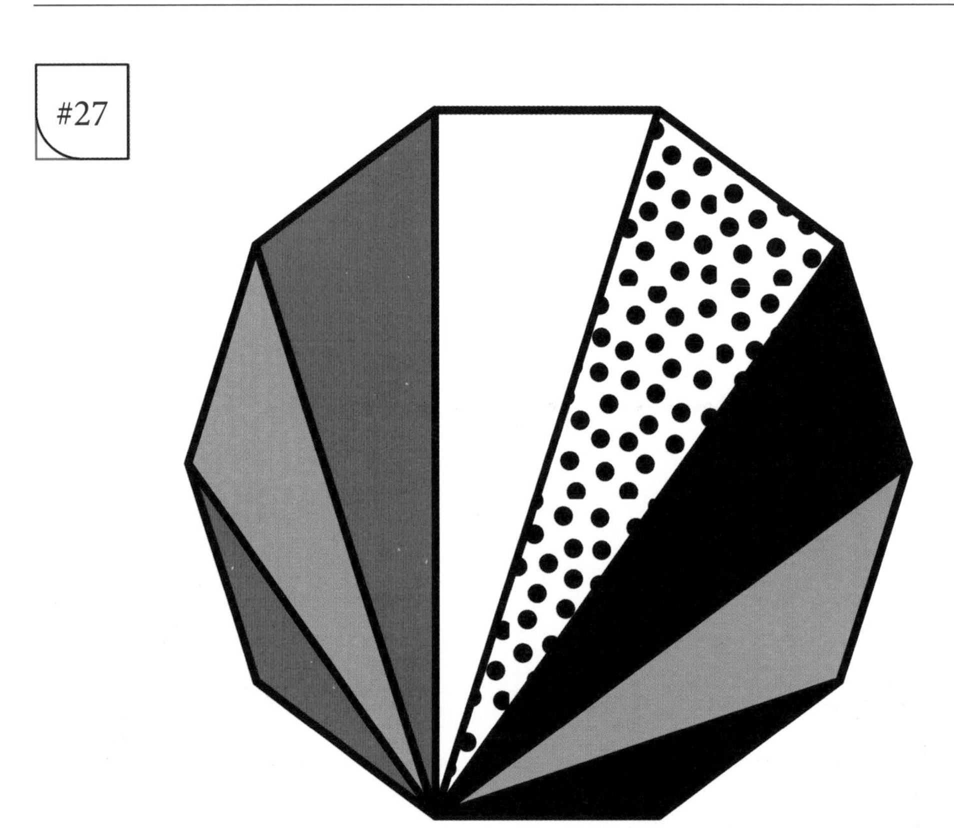

Prove that triangles with the same shading sum to one fifth of the total area of this regular decagon.

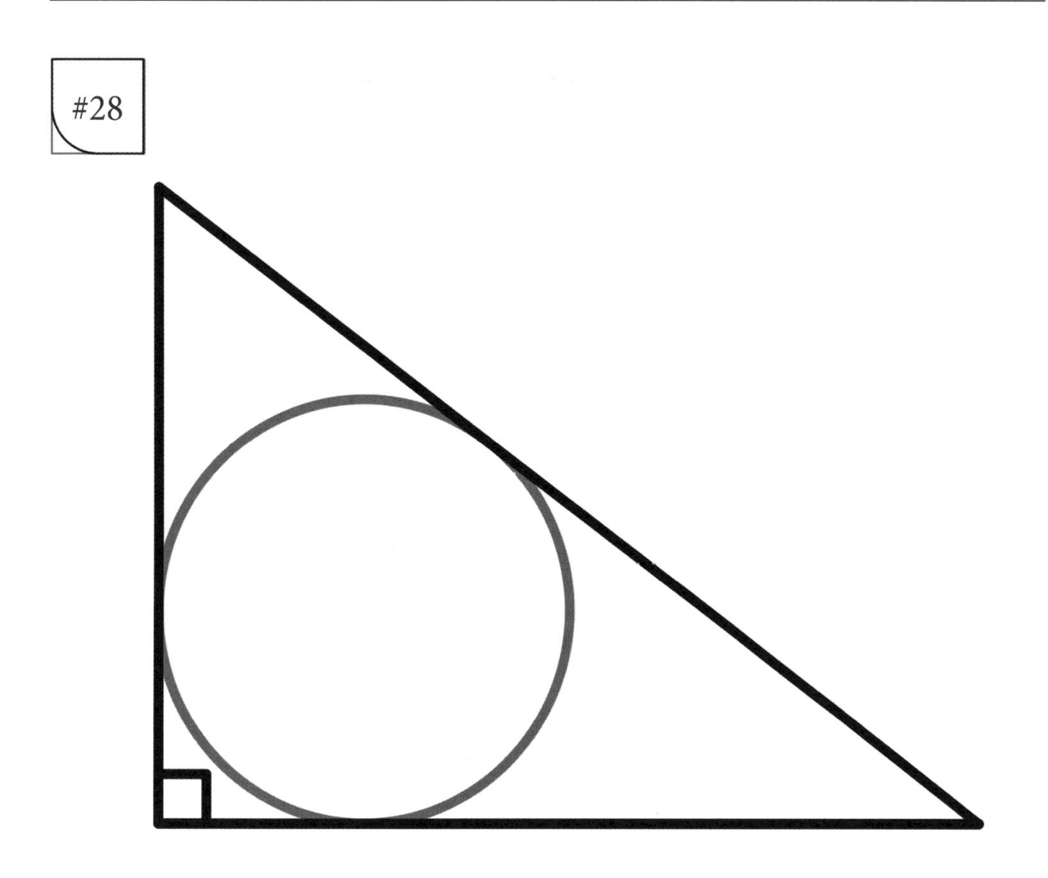

Prove that the area of a right angled triangle is equal to $s \times r$ where s is its semi-perimeter and r the radius of the incircle.

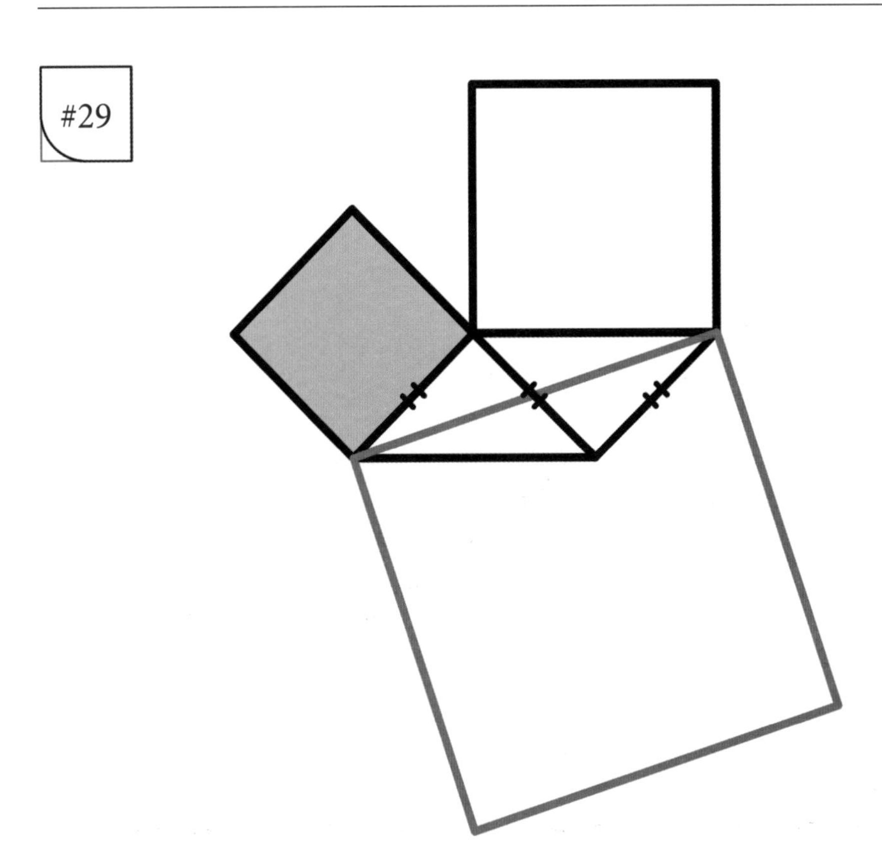

Three squares are constructed as shown. Prove that the area of the large red square is five times greater than the area of the shaded square.

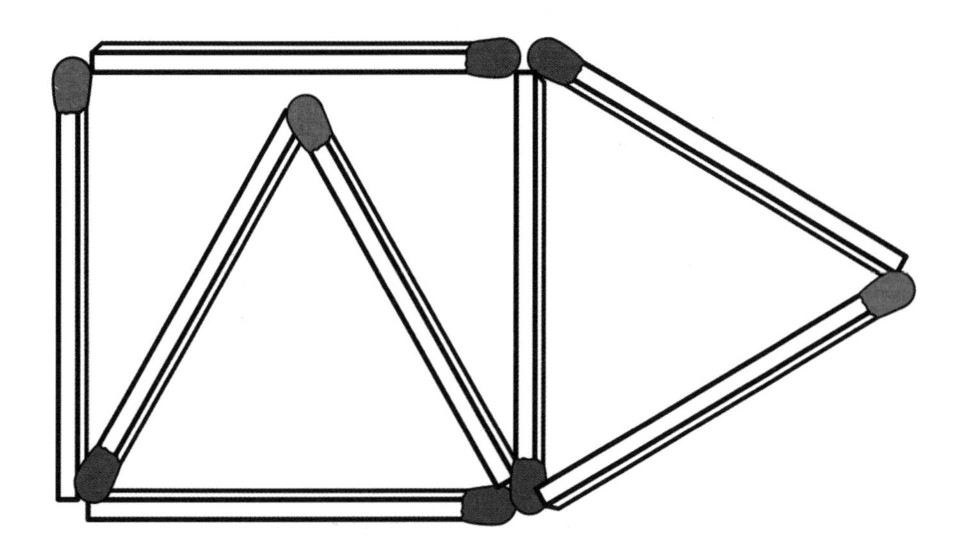

For the matchstick construction shown, prove that the three unused matchstick heads are collinear.

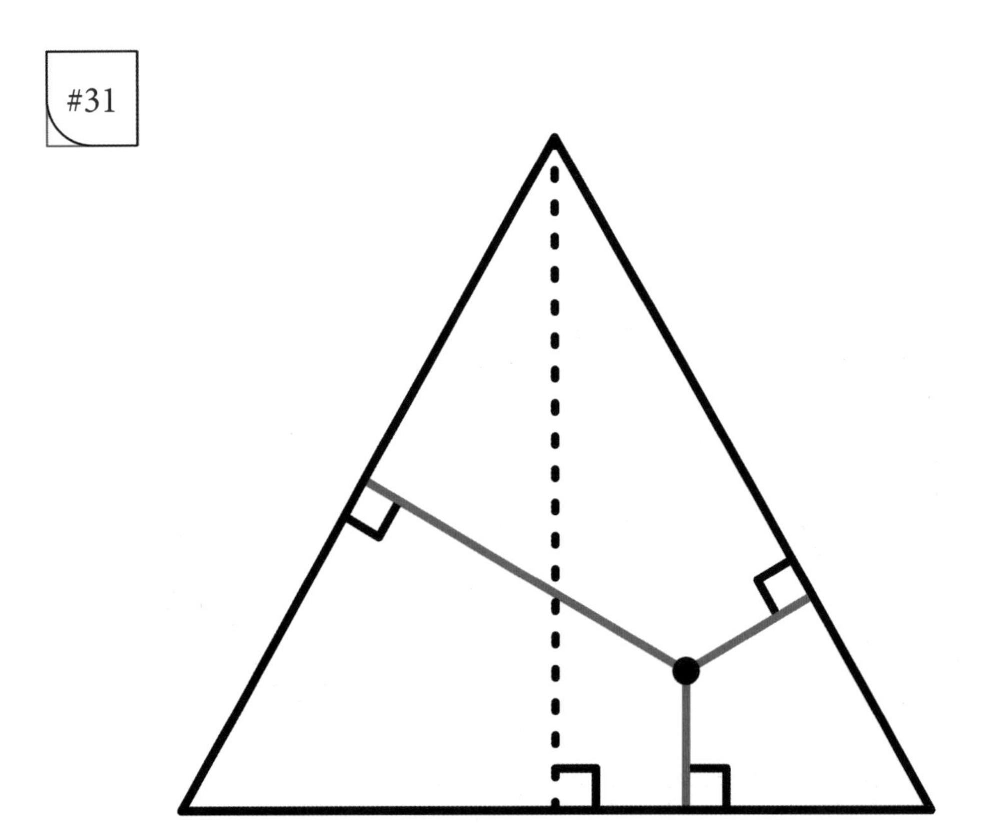

Prove that for any point inside an equilateral triangle, the sum of minimum distances to all 3 sides is equal to its height.

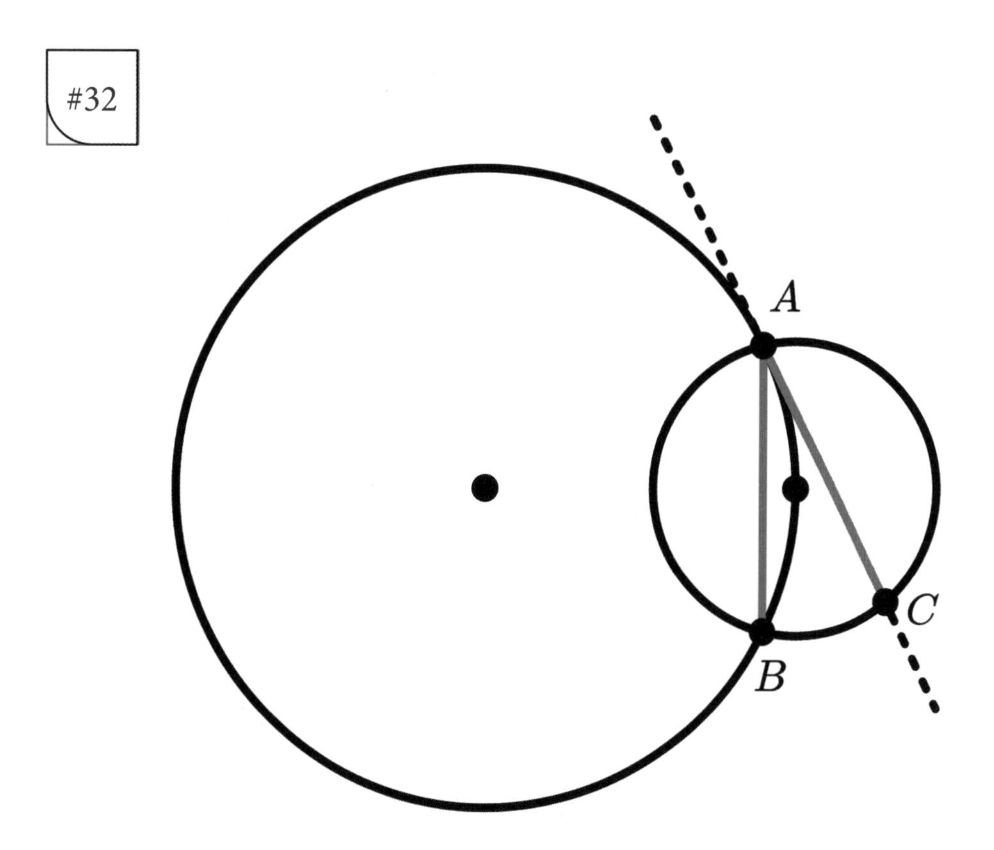

For two circles constructed such that one passes through the centre of the other, prove that the chord at their intersection, *AB*, and the chord made by tangent line *AC*, have equal length.

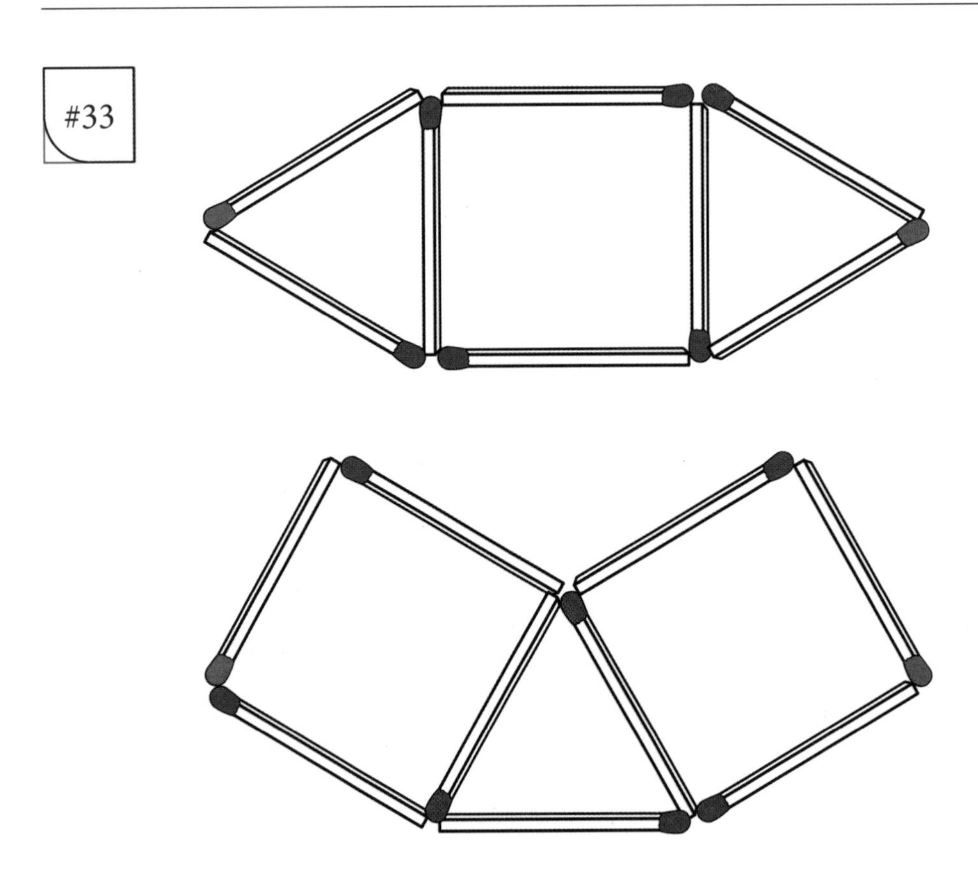

For the two matchstick constructions shown, prove that the straight lines between the unused matchstick heads on the outermost vertices in each diagram have the same length.

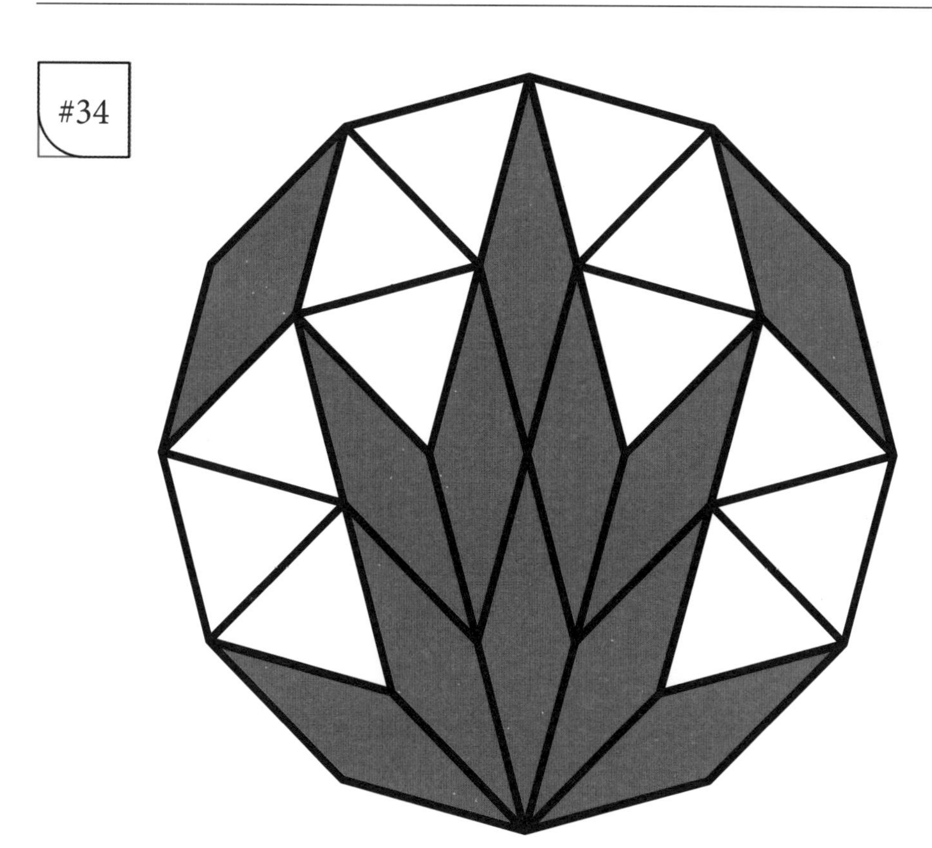

Prove that however you tessellate this regular dodecagon with these equilateral triangles and rhombi, you will always need twelve of each.

#35

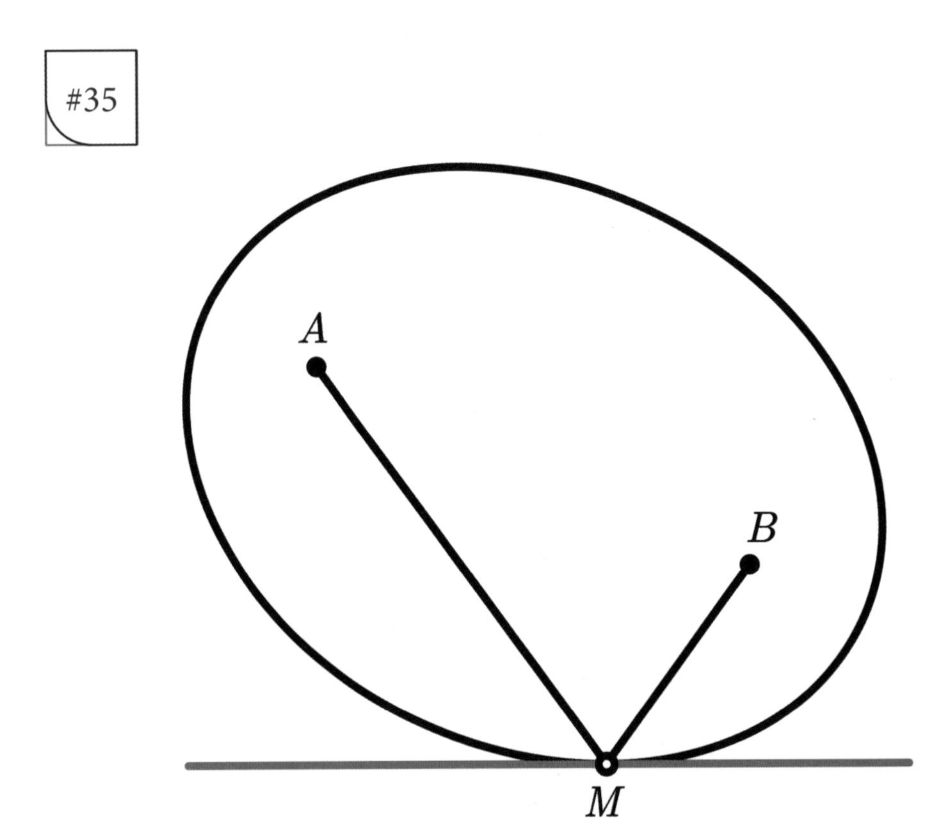

Prove that the shortest path from A to B when making a stop at M somewhere on the horizontal red line is such that the ellipse with focal points A and B going through M is tangent to the red line.

#36

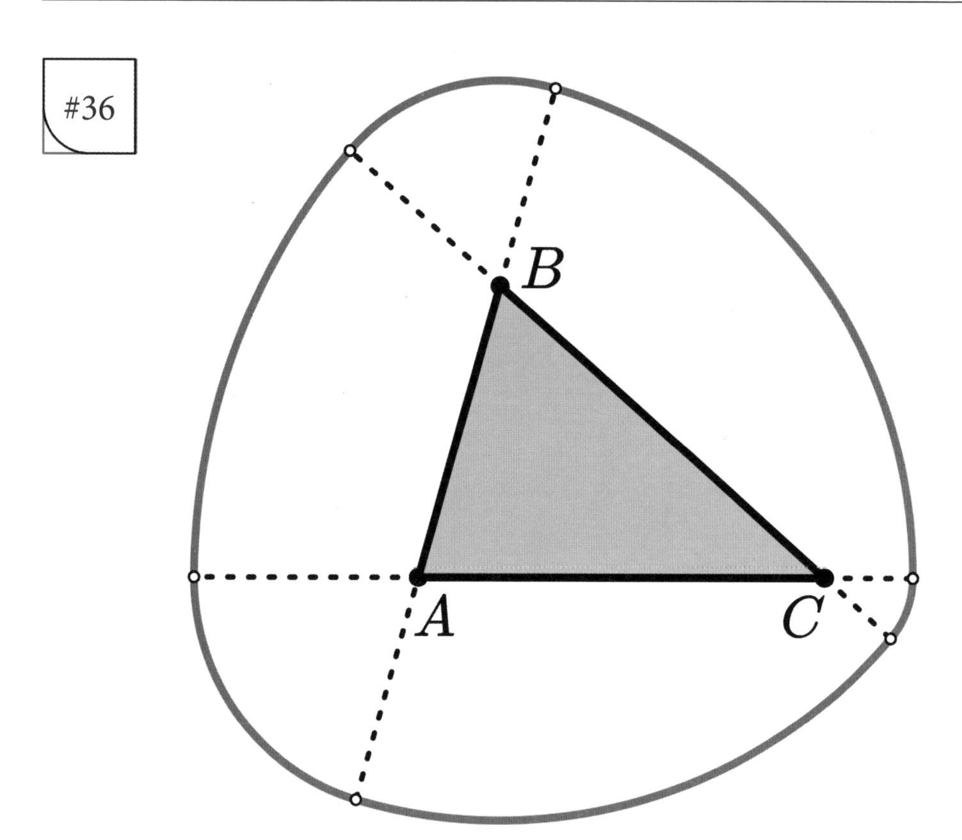

ΔABC is a general triangle. By extending the sides and drawing 6 arcs of circles with centres A, B, C, we can construct the curved red shape as shown - which has constant width d.

Prove that the circumference of the curved red shape is πd

Solutions and Answers

#25.

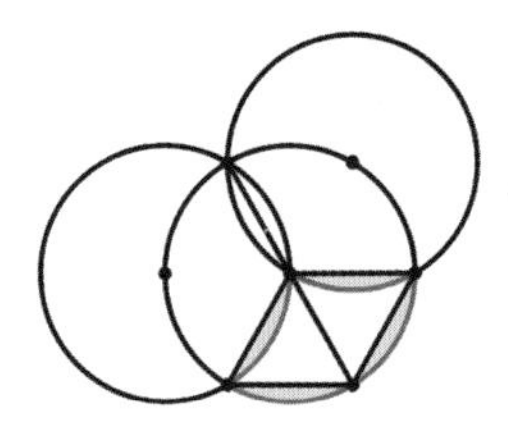

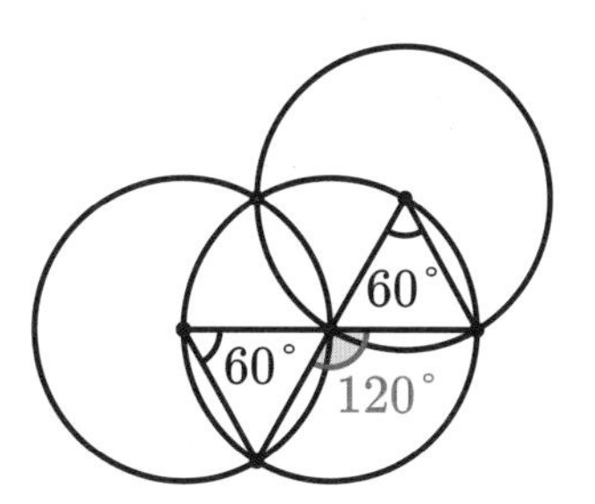

The shape can be visualised as a rhombus utilising only the radius.

$$\frac{120}{360}\pi r^2 - 2\left(\frac{60}{360}\pi r^2 - \frac{\sqrt{3}}{4} r^2\right) = \frac{\sqrt{3}}{2} r^2$$

#26.

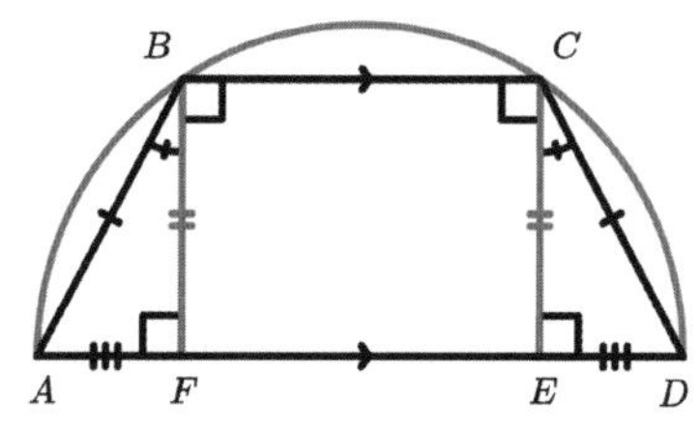

We know $\angle ACD = 90^\circ$, so $\angle ACB = \angle ECD = \angle ABF$

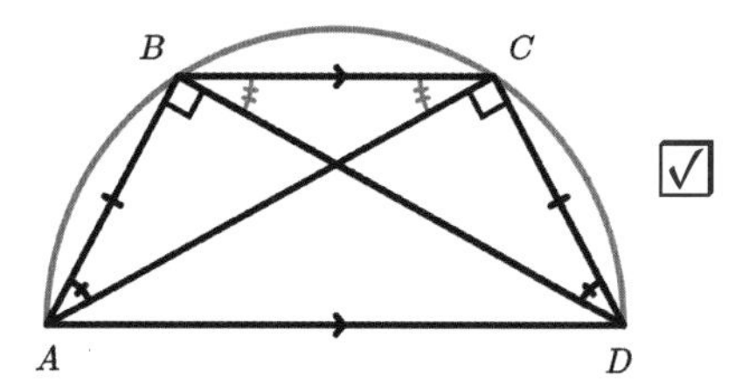

☑

#27.

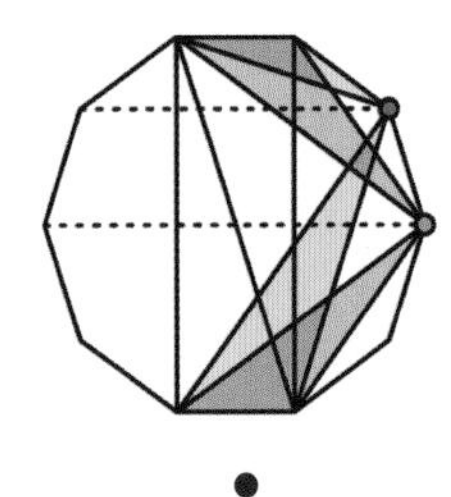

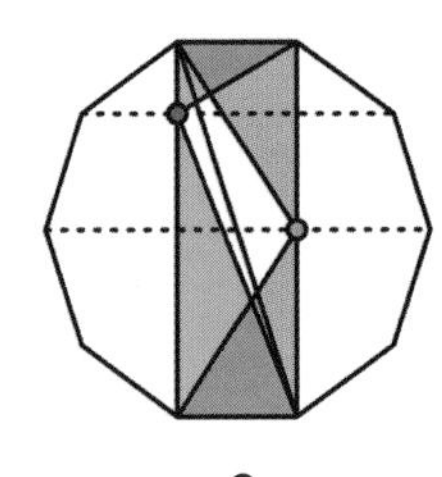

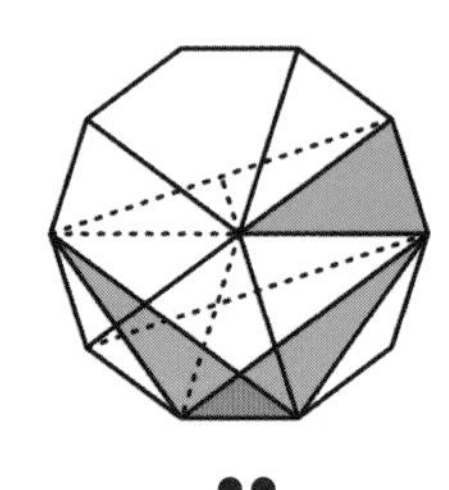

• By rearranging the triangles as shown, you can see that the total height of each combination is the same.

•• Split the decagon into fifths from the centre point. You can see that the bottom triangles are equal to one segment (a fifth) by shearing to the centre.

The other triangle is a quarter of the central rectangle, and also half of a fifth. Hence the remaining triangles (not shown) must sum to $\frac{2}{5}$ and there are two of each congruent type.

#28.

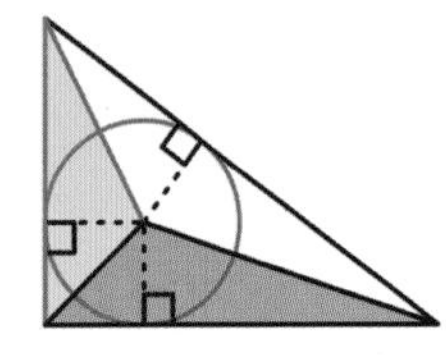

By splitting the total area A in three triangles where each base is a side (a, b, c) and each altitude is r, $A = \frac{1}{2}(ar + br + cr) = sr$.

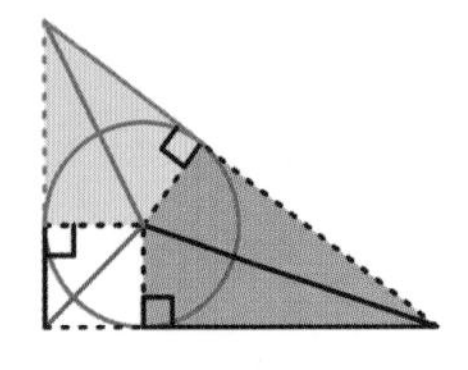

By splitting the total area A in six right triangles and and grouping them two by two, you see that their area sum up to $s \times r$ because the sum of the dashed legs is the semi-perimeter.

#29.

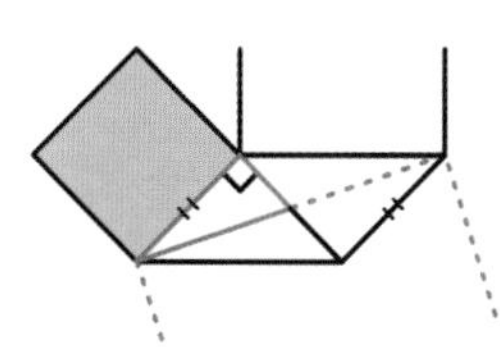

Let a be the side of the small shaded square and b the side of the largest square. Applying the Pythagorean theorem to the triangle formed from quarter of the parallelogram:

$$\left(\frac{b}{2}\right)^2 = a^2 + \left(\frac{a}{2}\right)^2$$

Multiply both sides by four to get $b^2 = 4a^2 + a^2 = 5a^2$.

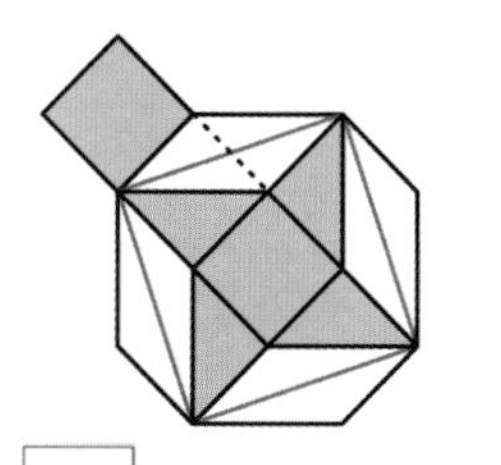

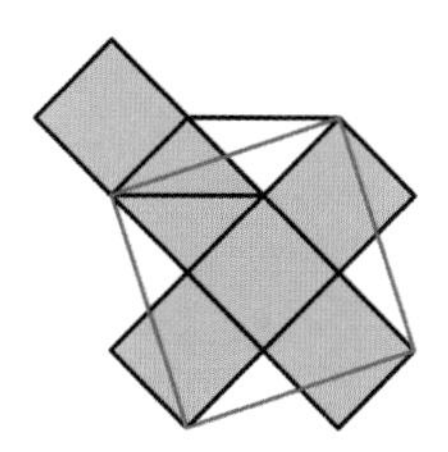

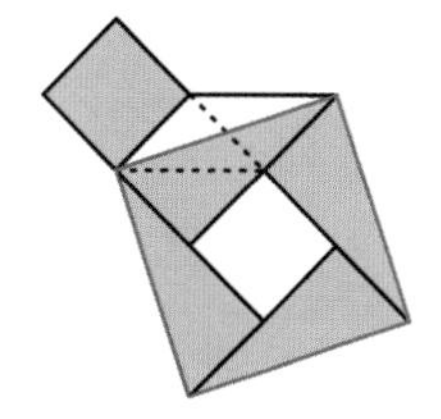

#30.

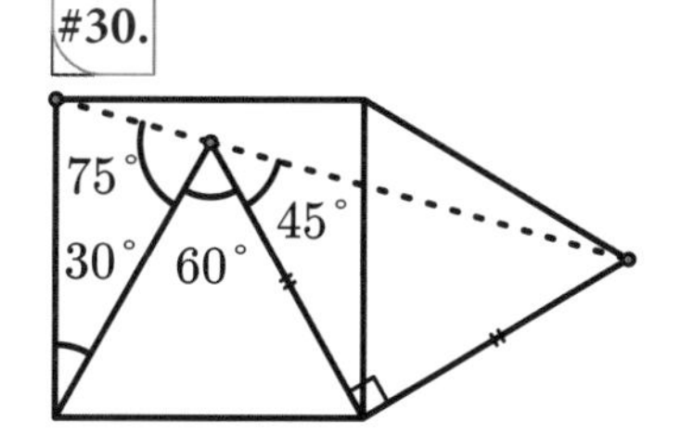

Using the angles in several equilateral, isosceles or right triangles you'll find the three angles adding up to a straight angle: 75 + 60 + 45 = 180°.

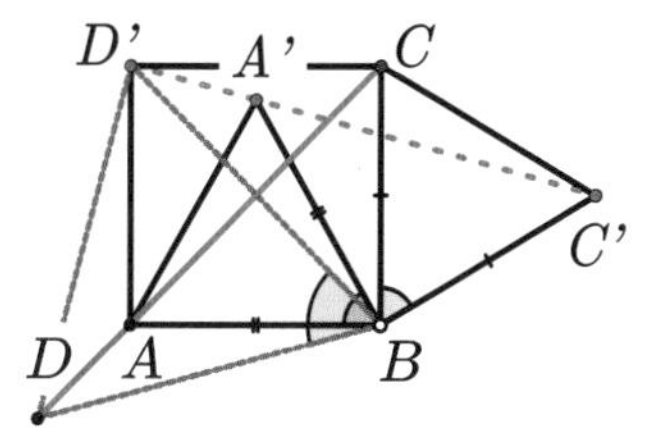

Let R be the 60° rotation with centre B (clockwise) and D the point such that $R(D) = D'$. The points D, A, C are collinear because they are equidistant from B and D' so the red line DC is the perpendicular bisector of $[BD']$. The rotated points D', A', C' are therefore also collinear.

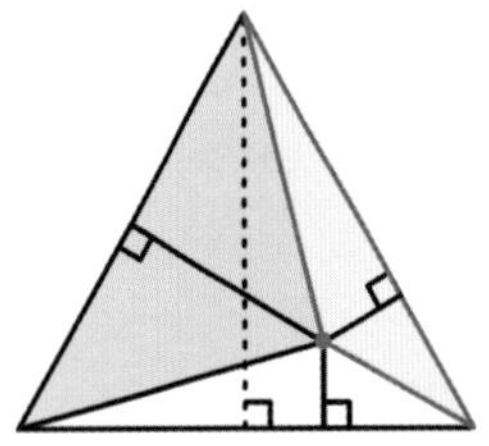

By splitting the triangle into three as shown, and calculate its area you can deduce that the combined heights must equal the altitude of the equilateral triangle using $A = \frac{bh}{2}$.

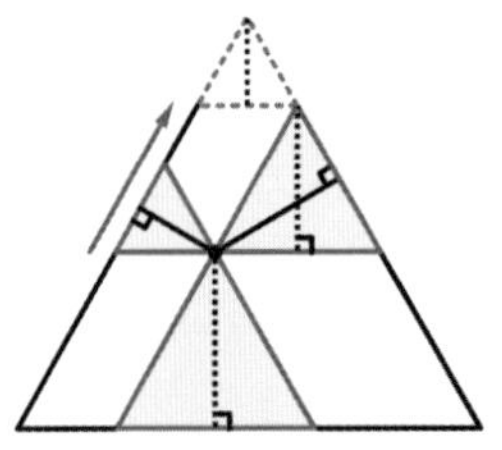

By using the three shaded equilateral triangles and moving one you get the result.

#32.

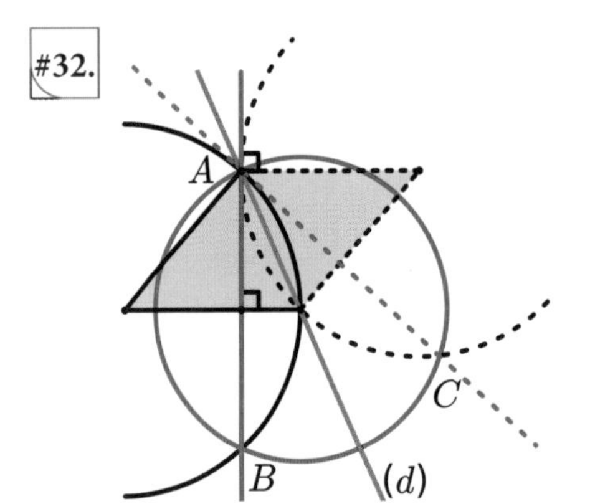

The tangent and the chord appear as reflections one of another along the line (d).

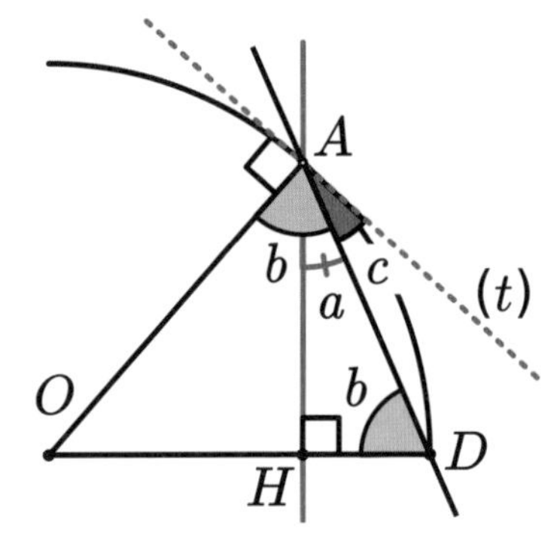

By proving a and c are equal, the chords must form two congruent isosceles triangles.

#33.

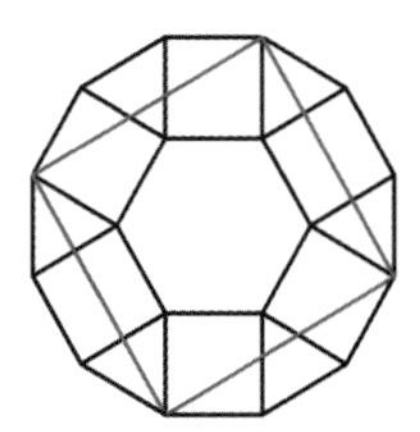

The two patterns appear as sides of a square in a regular dodecagon.

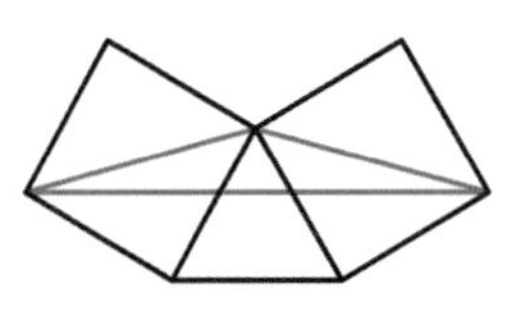

Say a match has unit length. For the square with two equilateral triangles the required length is

$$1 + 2 \times \frac{\sqrt{3}}{2} = 1 + \sqrt{3}.$$

For the other figure, using the cosine rule in this isosceles triangle with sides $\sqrt{2}$ and angle $45 + 60 + 45 = 150 = 180 - 30^{\circ}$:

$$x^2 = 4(1 + \cos(30)) = 4 - 2\sqrt{3} = \left(1 + \sqrt{3}\right)^2 \text{ so } x = 1 + \sqrt{3}$$

#34.

The area of the dodecagon is $6+3\sqrt{3}$. Adding halves (area of a rhombus) will not bring any $\sqrt{3}$. In order to have $3\sqrt{3}$ you need at least $3 \times 4 = 12$ triangles of area $\frac{\sqrt{3}}{4}$ (and 12 rhombi). If you put more triangles you will have fractions of $\sqrt{3}$ that cannot be cancelled by removing halves.

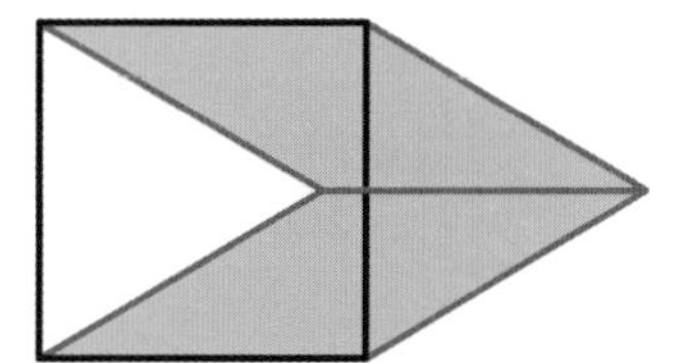

All segments have unit length. The area of a rhombus is $\frac{1}{2}$. The area of an equilateral triangle is $\frac{\sqrt{3}}{4}$. So the area of the dodecagon is $6 + 3\sqrt{3}$. If you add n rhombuses you will have to remove say m triangles so: $n \times \frac{1}{2} = m \times \frac{\sqrt{3}}{4}$. But that implies: $\frac{2n}{m} = \sqrt{3}$ which is impossible because $\sqrt{3}$ is irrational.

#35.

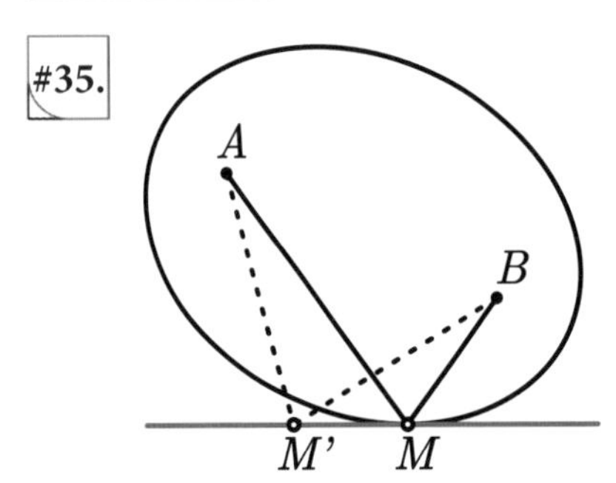

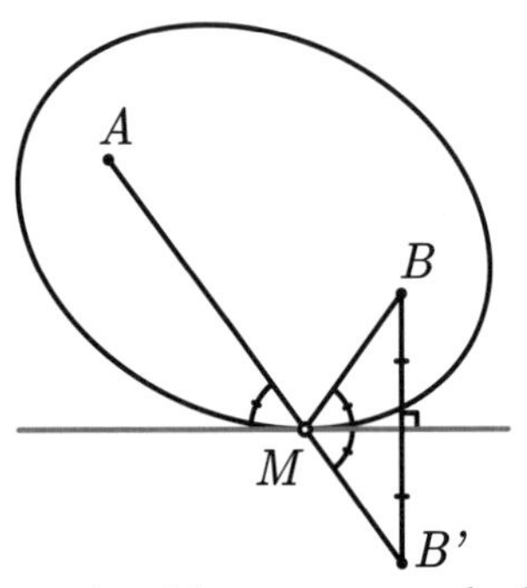

Consider the given ellipse tangent to the horizontal red line in M with focal points A and B. If you stop at another point M' on the red line the path will be longer because the points P on the ellipse satisfy $PA + PB = \text{cst}$ and the points on the outside $PA + PB > \text{cst}$.

Whatever point M you stop on the horizontal red line, $AM + MB = AM + MB'$ where B' is the reflection of B along the red line. In order to find the shortest path, A, M and B' must be collinear. In that case you have congruent angles as shown and the red line is therefore tangent to the ellipse going through M with A and B as focal points.

#36.

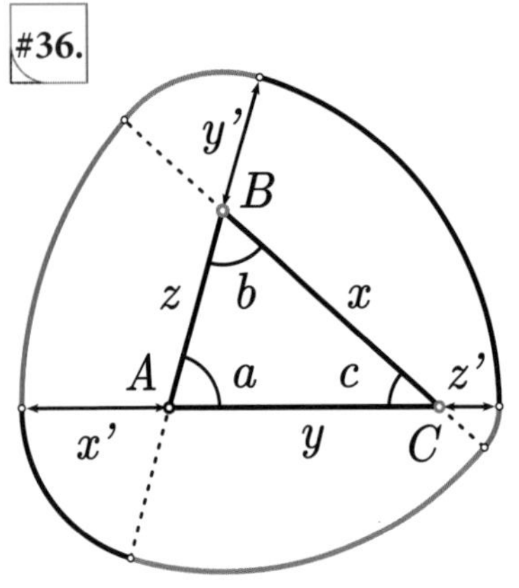

If angles a, b, c are in radians, the red part of the circumference is

$$z'c + (x + y')c = (x + y' + z')c = dc.$$

In the same way with the two other pair of arcs and adding them we get the perimeter:

$$P = da + db + dc = d(a + b + c) = \pi d$$

Writing three equations for d such as $d = x + y' + z'$ one can express x', y' and z' in terms of x, y, z and d and then calculate each arc length and add them to get the answer. A bit tedious.

What's the Area?

The aim of these puzzles is to find the area indicated. Any assumptions required to solve the puzzles are outlined in the text below each figure. In the solutions, the letter A will denote the area that the text is asking to calculate.

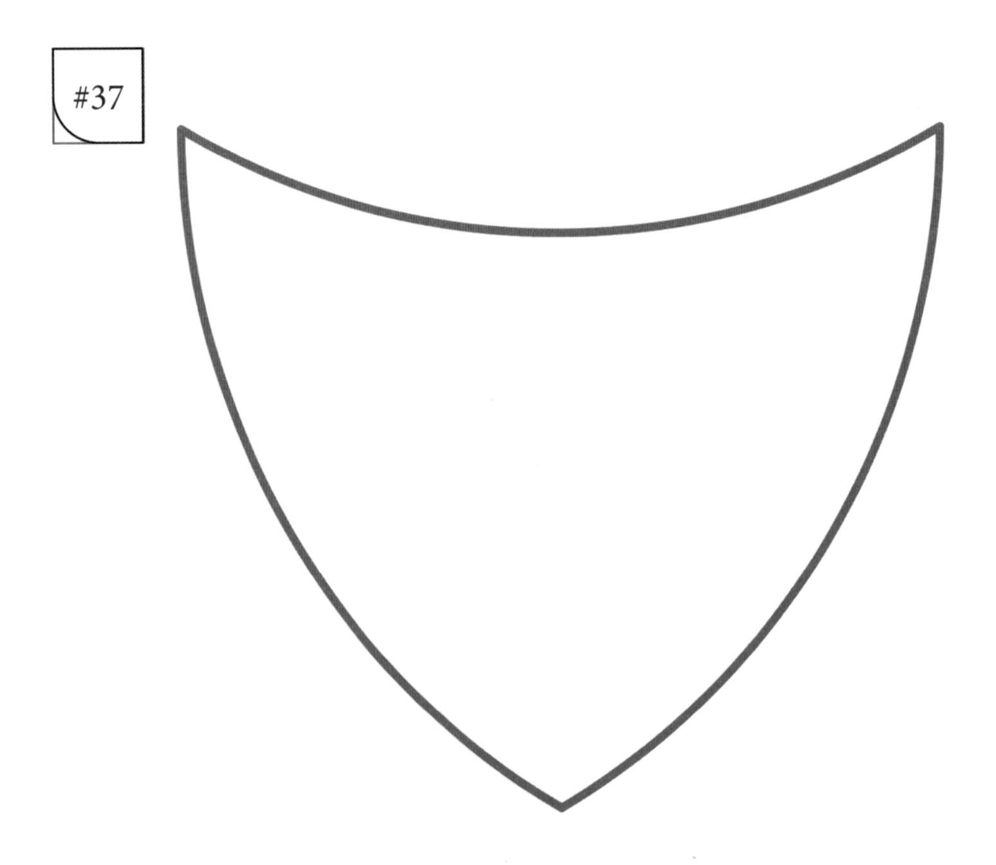

A shield-like shape is constructed using three 60° sectors of a unit circle with one arc inverted. Find the area of the shield.

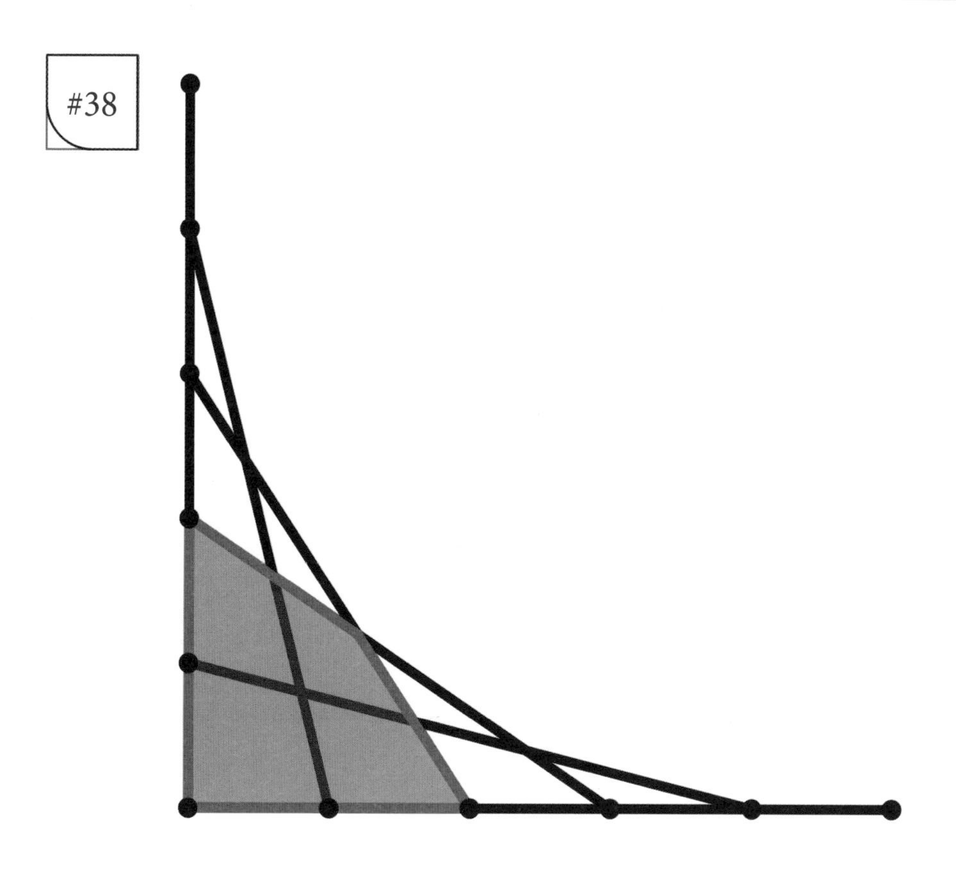

Two perpendicular line segments of length 5 cm with equally spaced points form a curve as shown.

Find the area of the red region.

#39

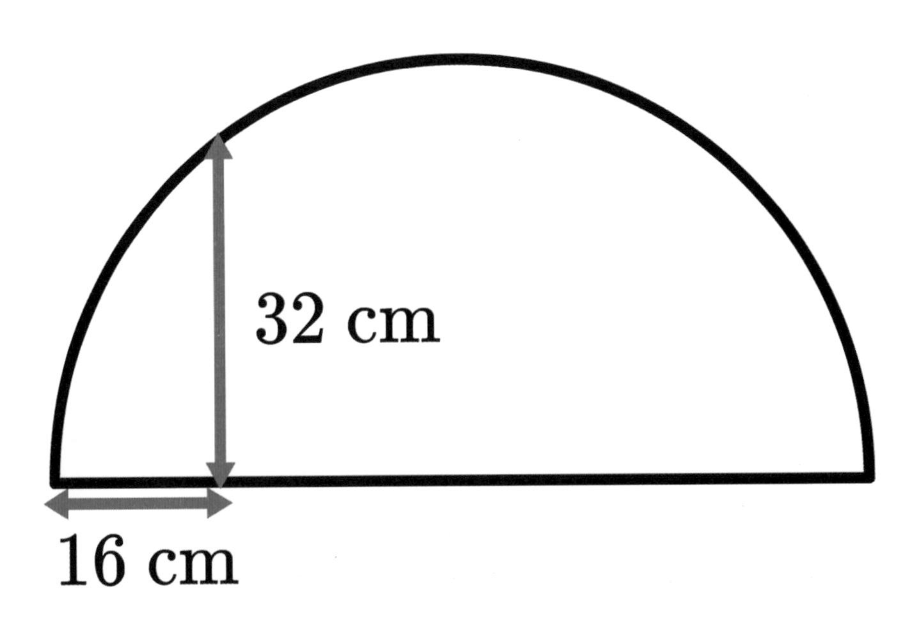

Find the area of the semicircle.

#40

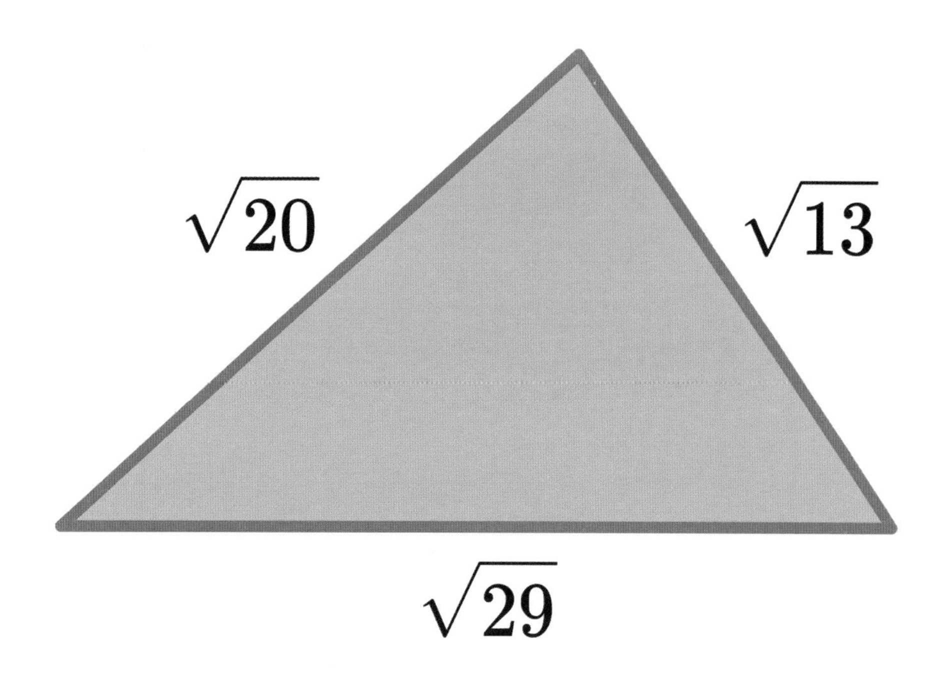

Find the area of this triangle.

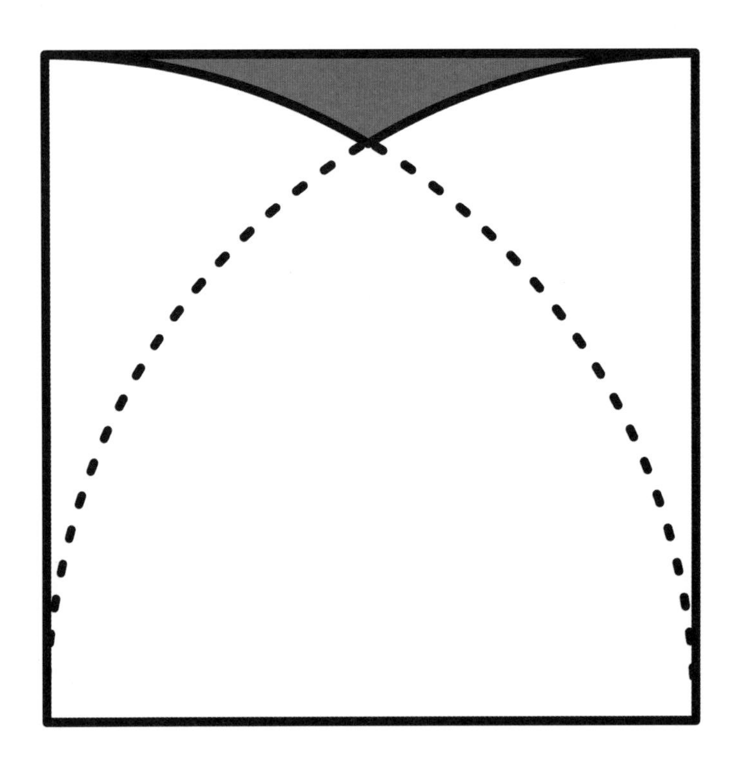

A square is constructed using two quadrants of a unit circle as shown. Find the area of the shaded region.

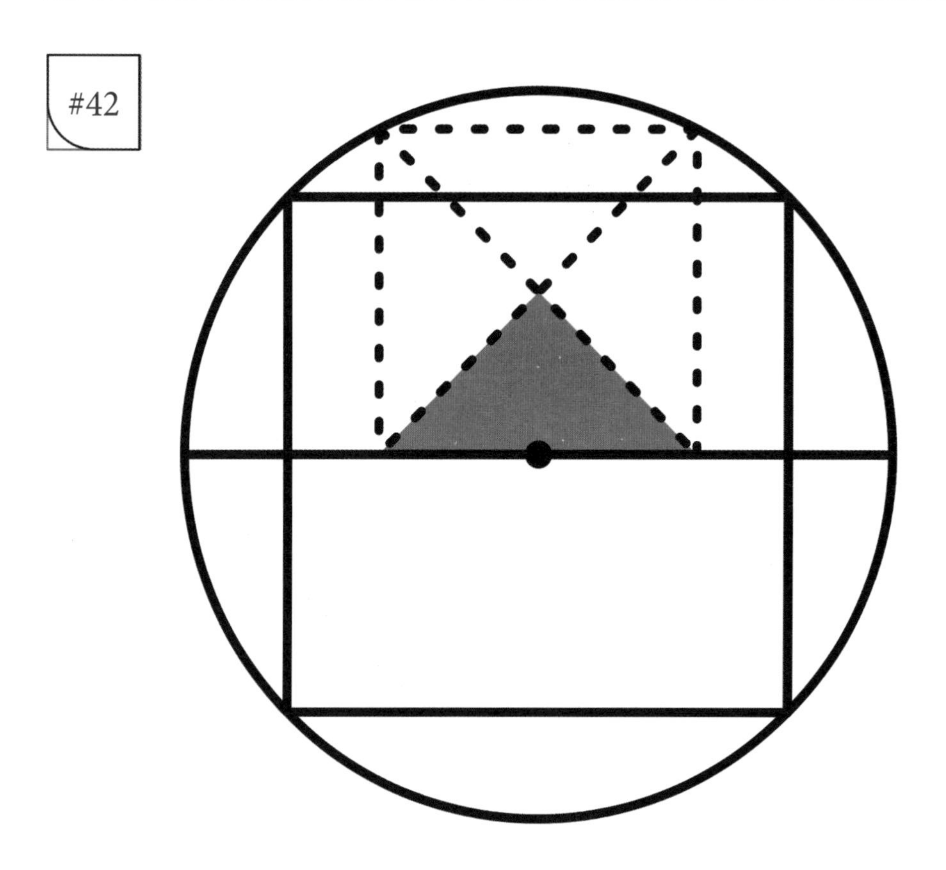

If the area of the triangle is 4 cm^2, find the area of the circumscribed square.

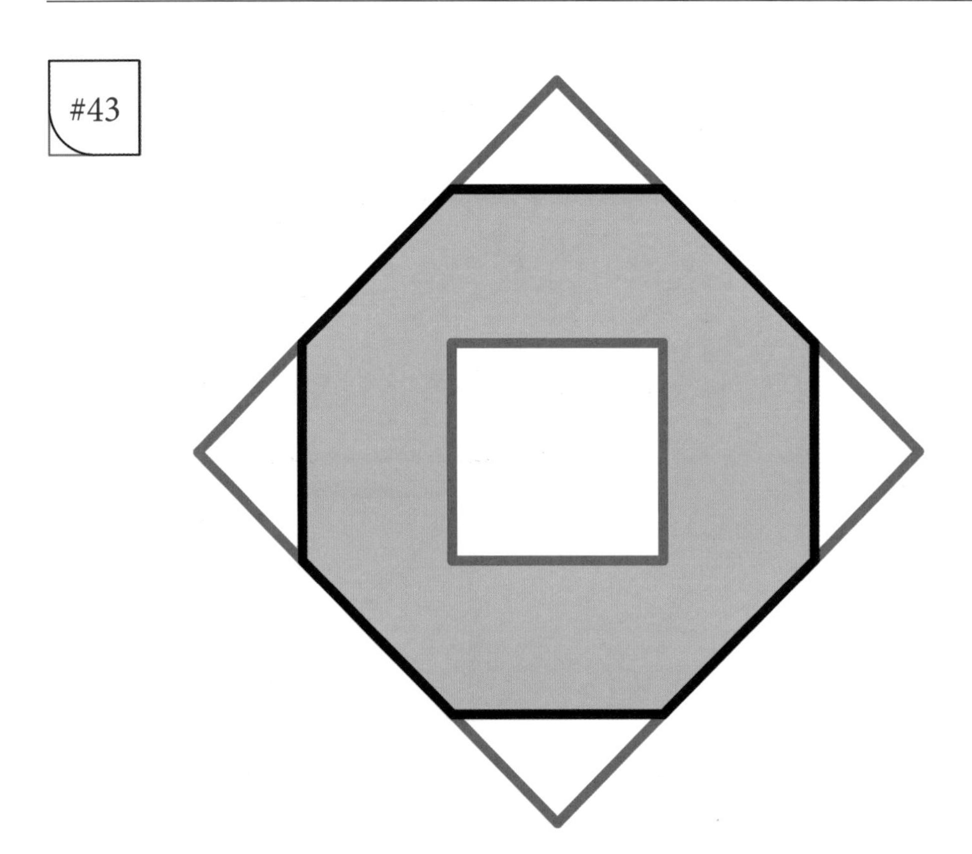

A concentric unit square and regular octagon are constructed inside a larger square as shown. The shapes have congruent side lengths. Find the area of the shaded region.

#44

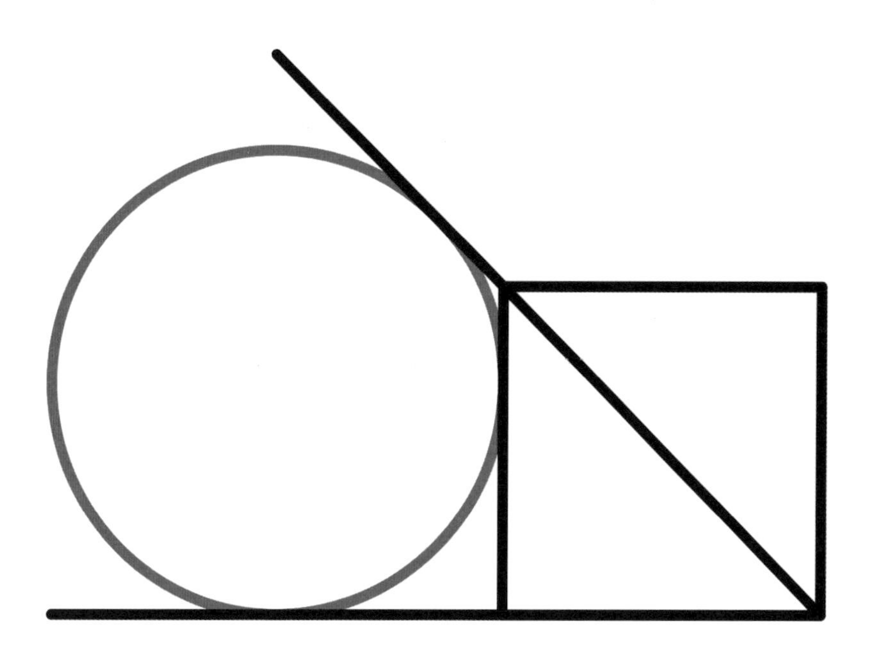

A square is constructed using three tangents to a unit circle as shown. What is the area of the square?

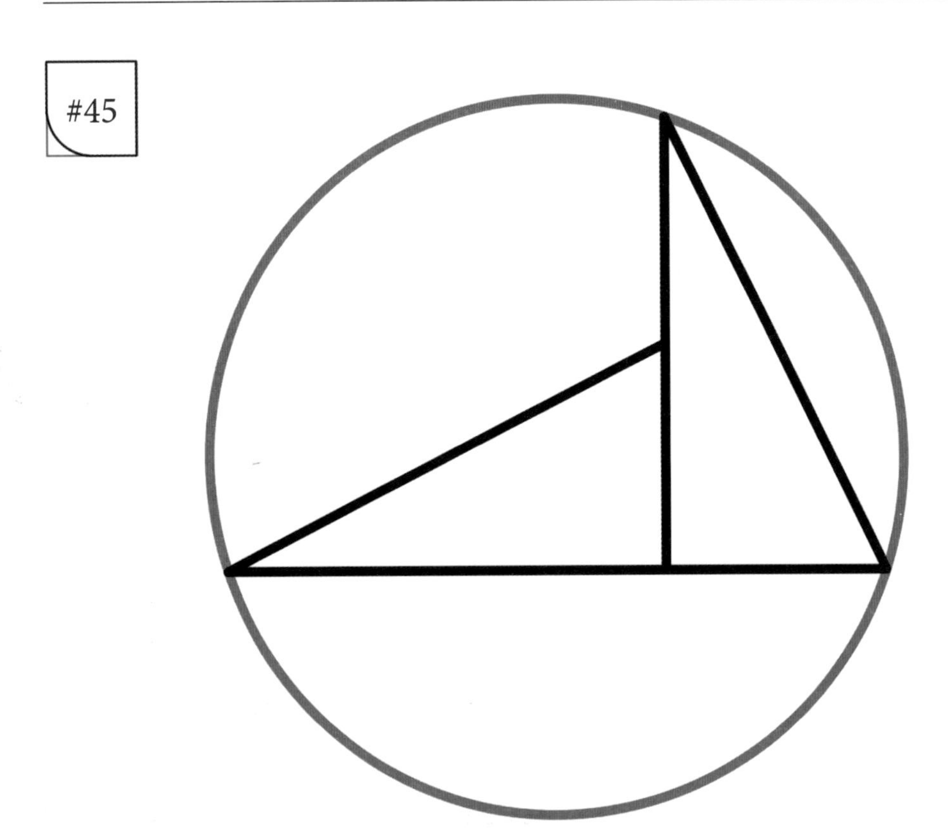

Two congruent right angled triangles with colinear touching sides and a leg length ratio of 2:1 touch a circle of radius 15cm at three vertices as shown. Find the area of one triangle.

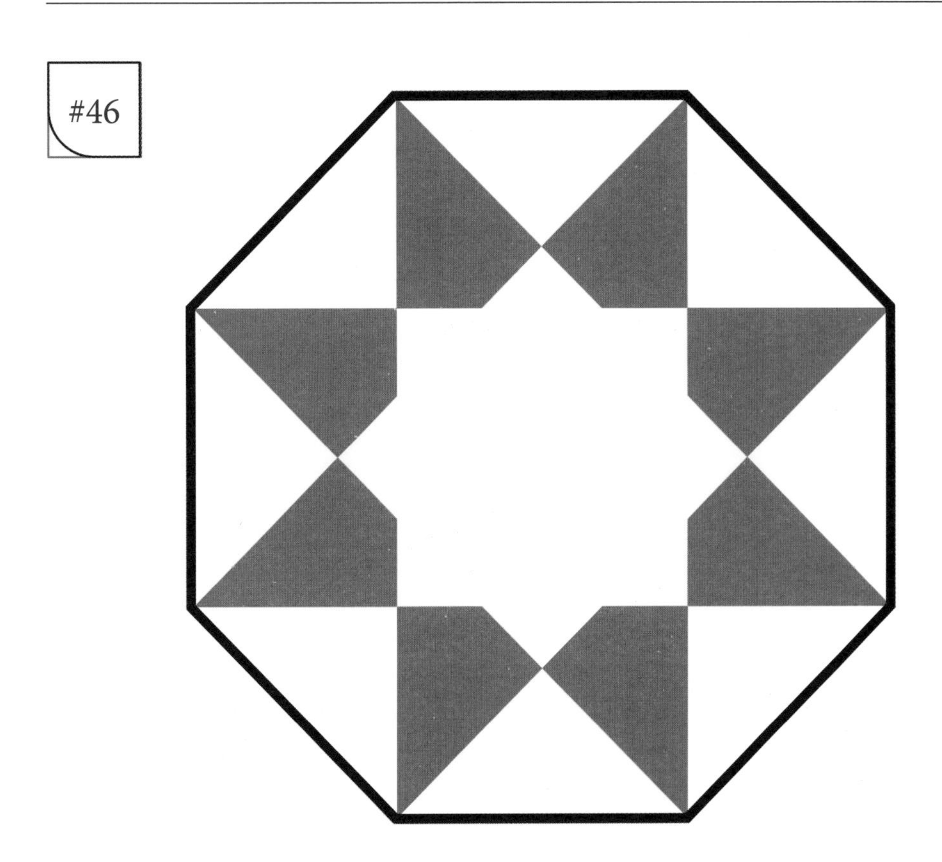

Find the area of this red star made using diagonals in a regular octagon of unit side length.

#47

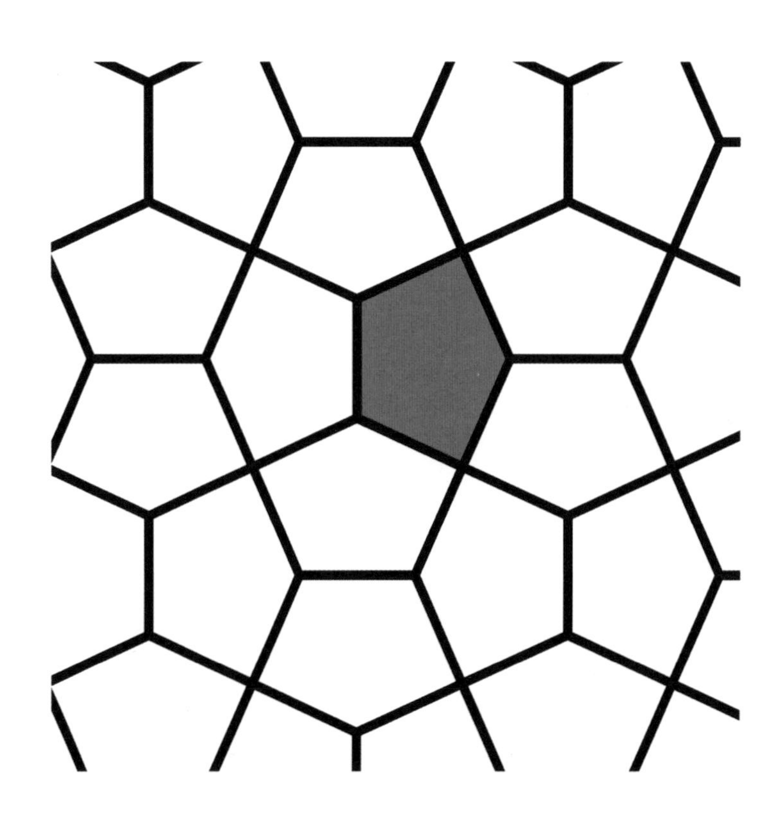

This tiling is made of congruent irregular pentagons with unit sides. Find the area of one tile.

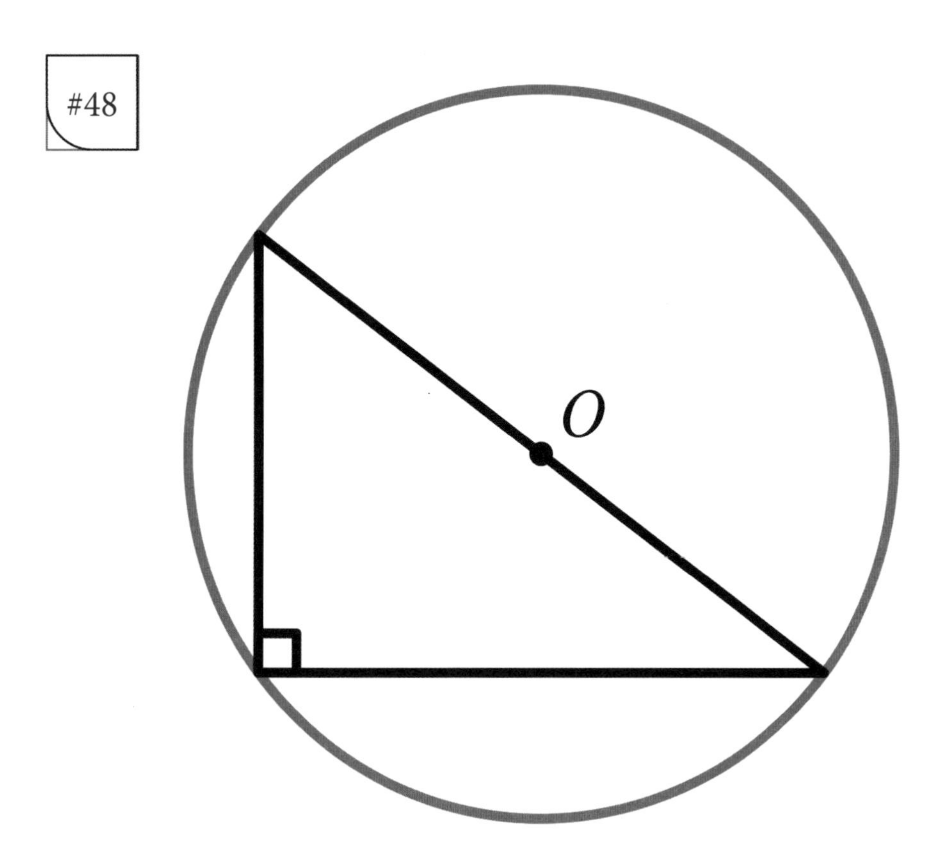

A right angled triangle with a perimeter of 72 cm and area of 216 cm^2 is circumscribed as shown. Find the area of the circle.

Solutions and Answers

#37. Answer: $A = \frac{\pi}{6}$ cm^2

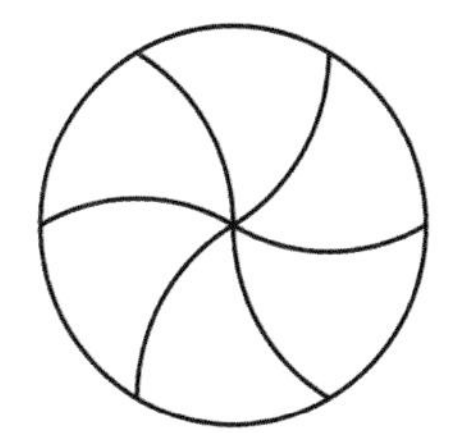

$A = \frac{1}{6}\pi r^2 = \frac{\pi}{6}$ cm^2

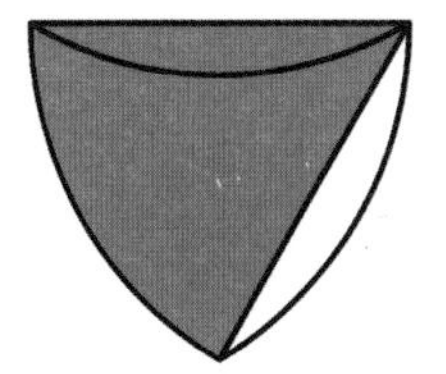

Rearranging one segment, one gets a 60° arc sector whose area is a sixth of a disk's. $A = \frac{\pi}{6}$ cm^2

#38. Answer: $A = \frac{12}{5}$ cm^2

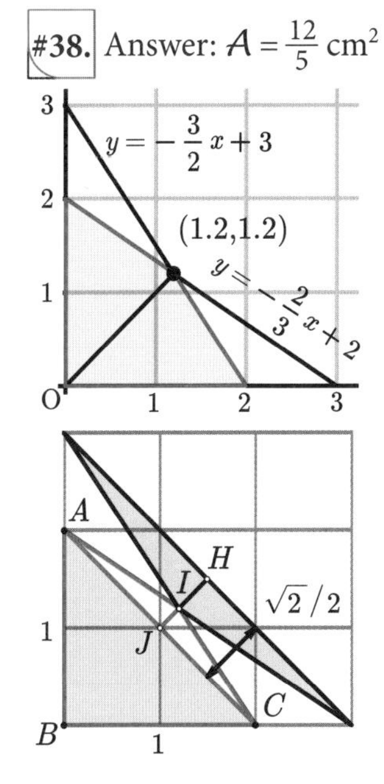

Using two equations of the straight lines $y = -\frac{2}{3}x + 2$ and $y = -\frac{3}{2}x + 3$, you solve for x:

$$-\frac{2}{3}x + 2 = -\frac{3}{2}x + 3 \Leftrightarrow \left(\frac{3}{2} - \frac{2}{3}\right)x = 1 \Leftrightarrow x = \frac{6}{5}.$$

Seen as two congruent triangles: $A = 2 \times \frac{6}{5} = \frac{12}{5}$ cm^2

The obtuse triangle with red sides and the grey are similar with ratio 2:3. Let $h = IJ$ and $h' = IH$ be the heights of ΔAIC and the grey triangle. From $h + h' = \frac{\sqrt{2}}{2}$ and $h' = \frac{3}{2}h$ we find $h = \frac{\sqrt{2}}{5}$. The area is that of ΔABC added to that of ΔAIC:

$$A = 2 + \sqrt{2} \times h \div 2 = 2 + \sqrt{2} \times \frac{\sqrt{2}}{5} = 2 + \frac{2}{5} = \frac{12}{5} \text{ cm}^2$$

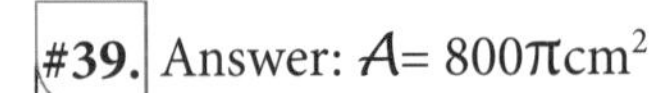

#39. Answer: $A = 800\pi \text{cm}^2$

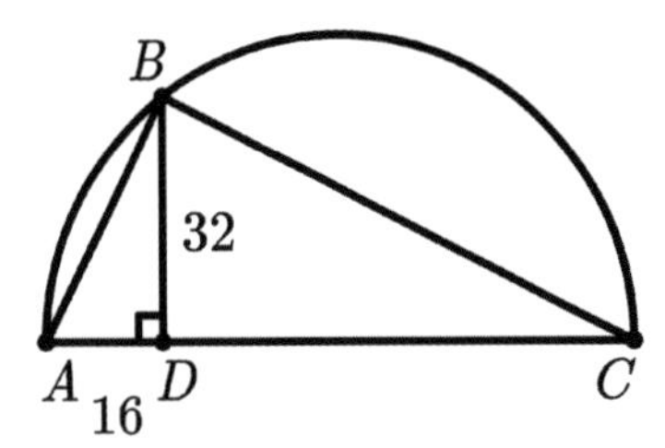

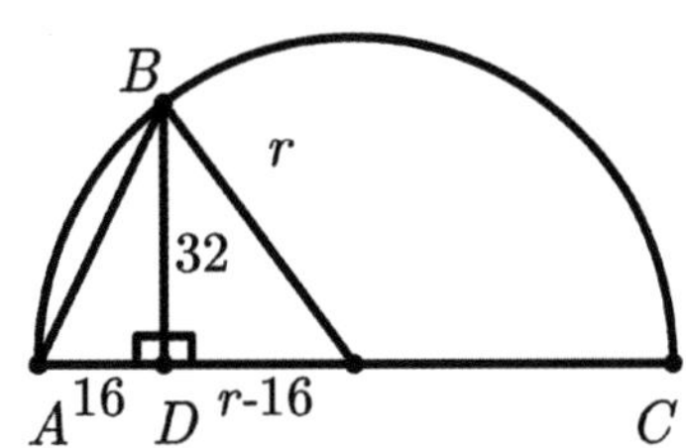

ABD and *BCD* are similar triangles. $16/32 = 0.5$ so *DC* must be 64 cm. Hence the radius is 40, and the area is $A = 800\pi \text{cm}^2$

$(r - 16)^2 + 32^2 = r^2$ so $1280 - 32r = 0$ so $r = 1280/32 = 40$, So $A = 800\pi \text{cm}^2$

#40. Answer: $A = 8 \text{cm}^2$

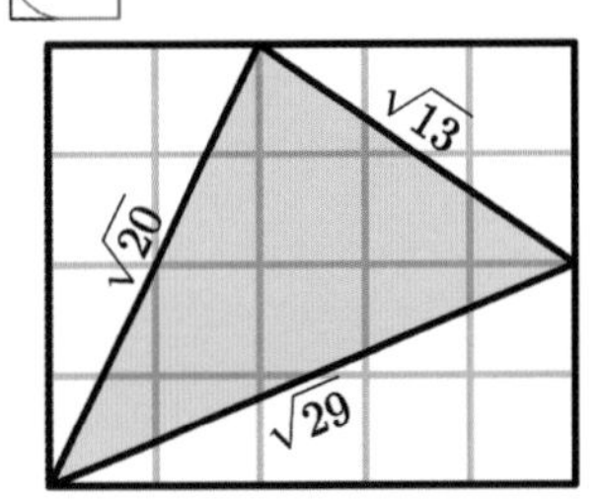

If you realise that 13, 20, 29 are sums of squares: $13 = 9 + 4$, $20 = 4 + 16$ and $29 = 25 + 4$ you can draw the rectangle shown and then the triangle inside.

$$A = 4 \times 5 - \tfrac{1}{2}(2 \times 3 + 2 \times 4 + 2 \times 5) = 20 - (3 + 4 + 5) = 8 \text{ cm}^2$$

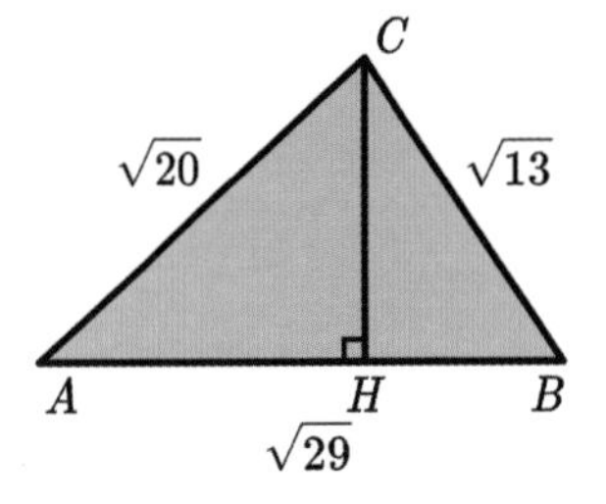

Calculating a dot product in two ways and using Pythagoras to find the altitude:

$\overrightarrow{AB} \cdot \overrightarrow{AC} = AH \times AB$ and also

$\overrightarrow{AB} . \overrightarrow{AC} = \tfrac{1}{2}(AB^2 + AC^2 - BC^2) = \tfrac{1}{2}(29 + 20 - 13) = 18.$

So $AH = 18/\sqrt{29}$. In ΔAHC: $CH^2 = AC^2 - AH^2 = \frac{256}{29}$.

Finally $A = \tfrac{1}{2} \times AB \times HC = \tfrac{1}{2} \times \sqrt{29} \times \frac{\sqrt{256}}{\sqrt{29}} = \frac{16}{2} = 8 \text{ cm}^2$

#41. Answer: $A = 1 - \frac{\sqrt{3}}{4} - \frac{\pi}{6}$ cm^2

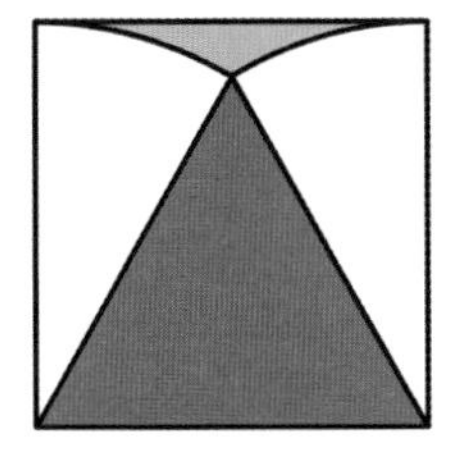

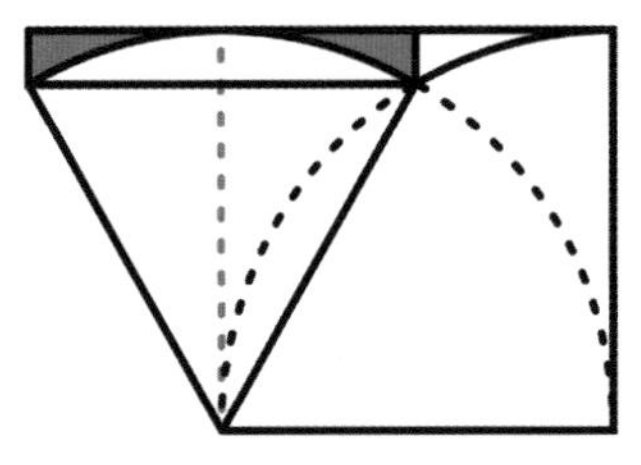

Using this dissection, A is a unit square minus an equilateral triangle and two 30° sectors of unit radius.

$A = 1 - \frac{\sqrt{3}}{4} - \frac{\pi}{6}$ cm^2

By reflecting half of the square vertically, the figure is transformed into an equilateral triangle with unit side lengths and a rectangle with unit length and height $(1 - \frac{\sqrt{3}}{2})$. Subtracting the segment from the rectangle gives $\left(1 - \frac{\sqrt{3}}{2}\right) - \left(\frac{\pi}{6} - \frac{\sqrt{3}}{4}\right) = A = 1 - \frac{\sqrt{3}}{4} - \frac{\pi}{6}$ cm^2

#42. Answer: $A = 40$ cm^2

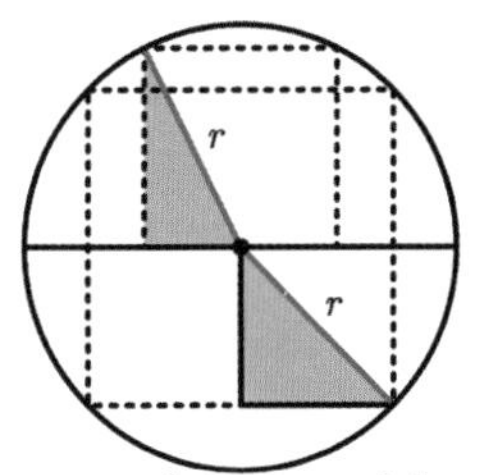

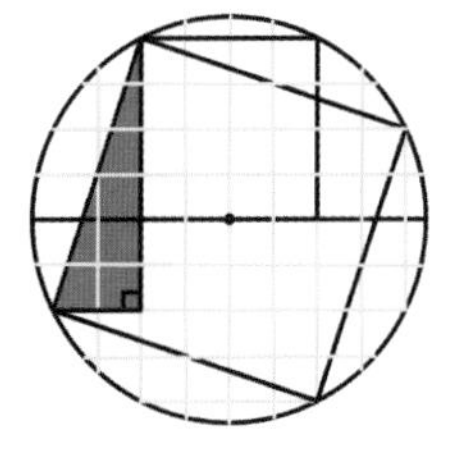

From the area of the triangle we deduce that the side of the small square is $\sqrt{4 \times 4} = 4$. The radius r of the circle satisfies then $r^2 = 4^2 + 2^2 = 20$. So the area A of the big square is $A = 2r^2 = 40$ cm^2

In this arrangement, you can see that one square is $\frac{2}{5}$ the size of the other.

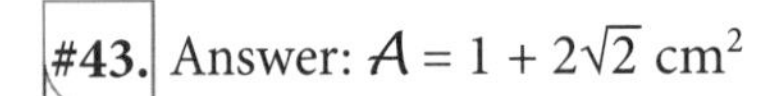

#43. Answer: $A = 1 + 2\sqrt{2}$ cm^2

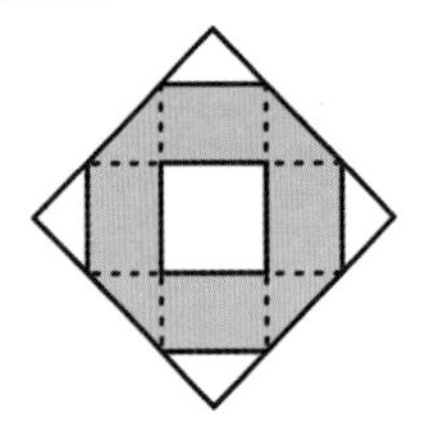

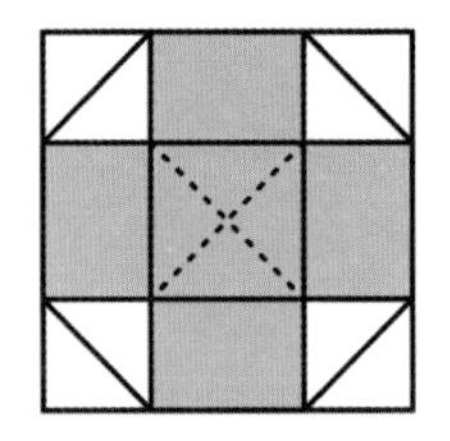

Rearranging the isosceles right triangles to make this cross we get:

$$A = (1 + \sqrt{2})^2 - 4(\tfrac{\sqrt{2}}{2})^2 = 1 + 2\sqrt{2} \text{ cm}^2$$

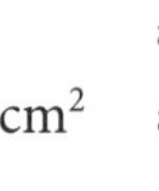

The central square has unit side. The shaded area can be seen as four rectangles $1 \times \frac{\sqrt{2}}{2}$ and four right isosceles with legs $\frac{\sqrt{2}}{2}$.

$$A = 2\sqrt{2} + 4\tfrac{1}{2}(\tfrac{\sqrt{2}}{2})^2 = 1 + 2\sqrt{2} \text{ cm}^2$$

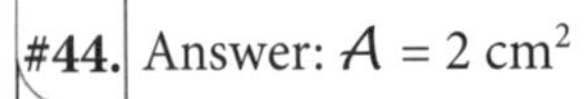

#44. Answer: $A = 2$ cm^2

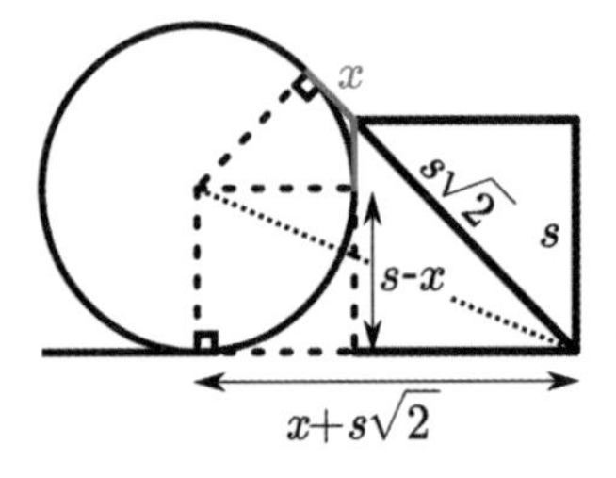

$s - x = 1 \Leftrightarrow x = s - 1$

$x + \sqrt{2}s = s + 1$

$\sqrt{2}s = 2$

$A = s^2 = 2$ cm^2

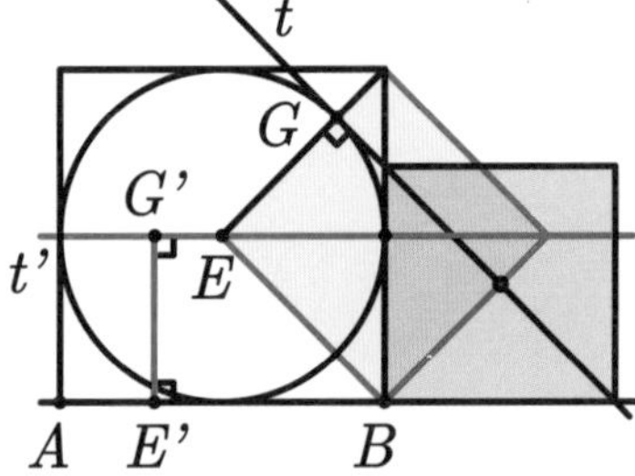

Let R be the 45° rotation with centre B. $R(E) = E' \in [AB]$ and $R([EG]) = [E'G']$ with $EG = E'G'$ = radius and $(E'G') \perp (AB)$. Looking at the tangent, $R(t) = t'$ goes through the centre and is parallel to (AB). So the image of the grey square is the red one whose side therefore measure $\sqrt{2}$, so the area of both red and grey squares is $A = 2$ cm^2

#45. Answer: $A = 90\text{ cm}^2$

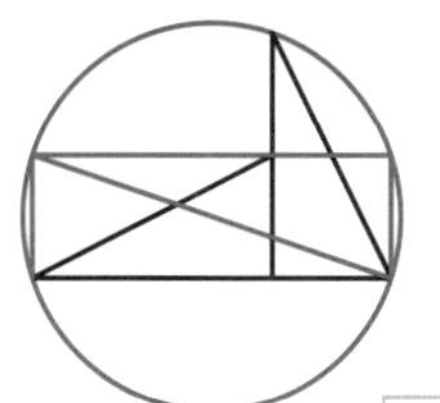

The large red triangle has legs of ratio 3:1 so its hypotenuse (which is the circle diameter) $= \sqrt{10s^2} = s\sqrt{10} = 30$ so the base of the original triangles is $3\sqrt{10}$, the height is $6\sqrt{10}$, so $A = 90\text{ cm}^2$

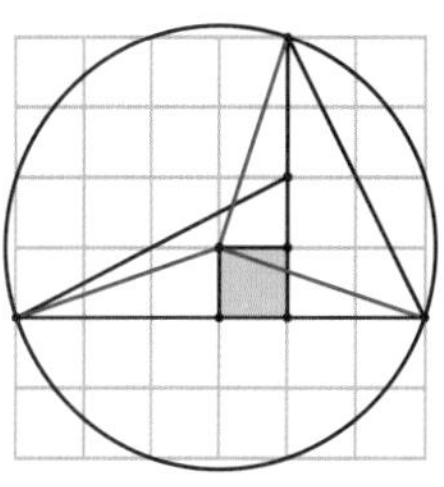

Say sides are 1 and 2 (so the triangles have unit area).

$(1) : r^2 = y^2 + \left(\frac{3}{2}\right)^2$ and

$(2) : r^2 = (2 - y)^2 + \left(\frac{1}{2}\right)^2$ so $(1) - (2) \Rightarrow y = \frac{1}{2}$. $r^2 = \frac{5}{2}$.

Enlarge to get a radius of 15 cm: $\frac{5}{2} \times k = 15^2 \Rightarrow k = 90$.

Area of one triangle is $A = 90\text{ cm}^2$.

#46. Answer: $A = 4(\sqrt{2} - 1)\text{ cm}^2$

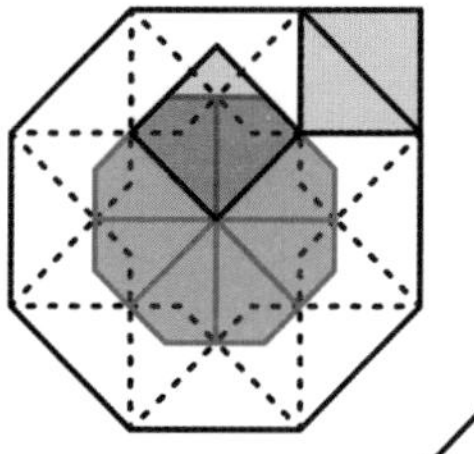

Because the grey shape is a square, each kite has the area of a red rectangle with sides $\frac{\sqrt{2}}{2}$ and $(1 - \frac{\sqrt{2}}{2})$.

So $A = 8 \times \frac{\sqrt{2}}{2} \times \left(1 - \frac{\sqrt{2}}{2}\right) = 4(\sqrt{2} - 1)\text{ cm}^2$

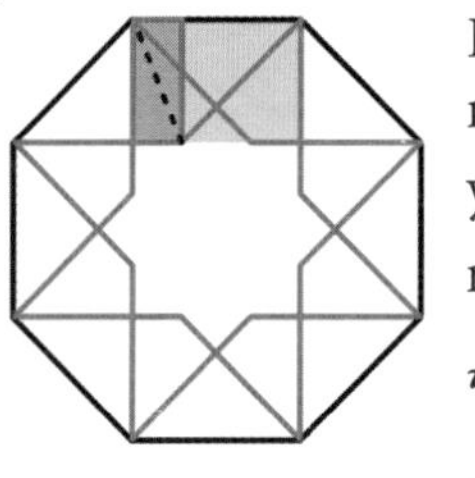

Rearrange the kites in the centre to form a smaller regular octagon. Using the grey congruent squares you see that the ratio of lengths is $1 : 1 + \frac{\sqrt{2}}{2}$ so the red area is $\left(1 - \frac{\sqrt{2}}{2}\right)^2 = \left(\frac{3}{2}\right) + \sqrt{2}$ smaller.

$$A = \frac{(1+\sqrt{2})^2 - 1}{\frac{3}{2} + \sqrt{2}} = \left(2 + 2\sqrt{2}\right)\left(\frac{3}{2} - \sqrt{2}\right) \div \left(\frac{9}{4} - 2\right)$$

$= 4(\sqrt{2} - 1)\text{ cm}^2$

#47. Answer: $A = 1 + \frac{\sqrt{7}}{4}$ cm^2

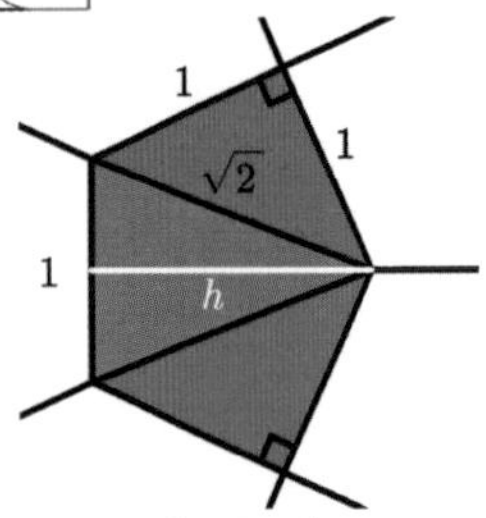

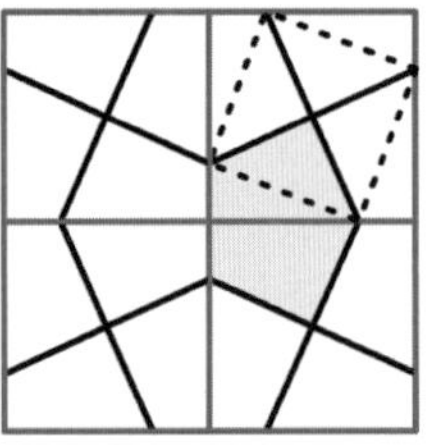

Using the Pythagorean theorem in one half of the black triangle: $h^2 = 2 - \frac{1}{4} = \frac{7}{4}$ so: $h = \frac{\sqrt{7}}{2}$. The two red triangles' areas sum to that of a unit square, so:

$A = 1^2 + h \times 1 \div 2 = 1 + \frac{\sqrt{7}}{4}$ cm^2

$h^2 + \frac{1}{4} = 2 \Rightarrow = h^2 = \frac{7}{4} \Rightarrow h = \frac{\sqrt{7}}{2}$. The area of the pentagon is half the area of a square, so:

$A = \frac{1}{2}(\frac{1}{2} + \frac{\sqrt{7}}{2}) = \frac{1}{2}(2 + \frac{\sqrt{7}}{2}) = 1 + \frac{\sqrt{7}}{4}$ cm^2

#48. Answer: $A = 225\pi$ cm^2

Let s be the semiperimeter of the triangle, and r the inradius.

$r = \frac{A}{s} = \frac{216}{36} = 6$. Using the formula[1]

$s - r = c$ we get: $36 - 6 = 30$

$A = \pi(\frac{30}{2})^2 = 225\pi$ cm^2

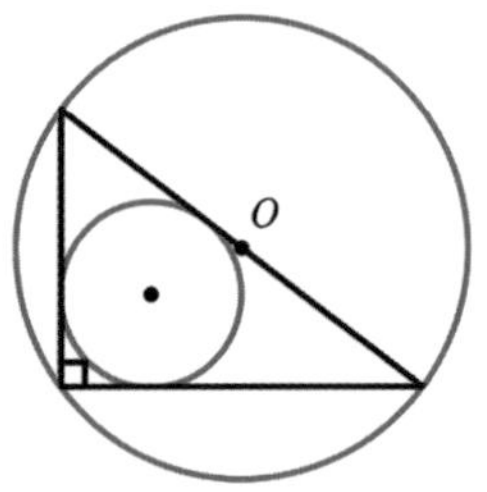

A second algebraic solution can be found online at the url below.

[1] The formula s − r = c is too lengthy to derive here, and is a fun challenge in itself. However a full derivation can be found online at https//www.tarquingroup.com/geomsnackssolution.

Sangaku

Sangaku puzzles are geometry problems that were painted on tablets inside Japanese temples between the 17th and 19th Century. The five puzzles provided in this book are based on some of these diagrams. Any assumptions required to solve the puzzles are indicated in the text below each figure.

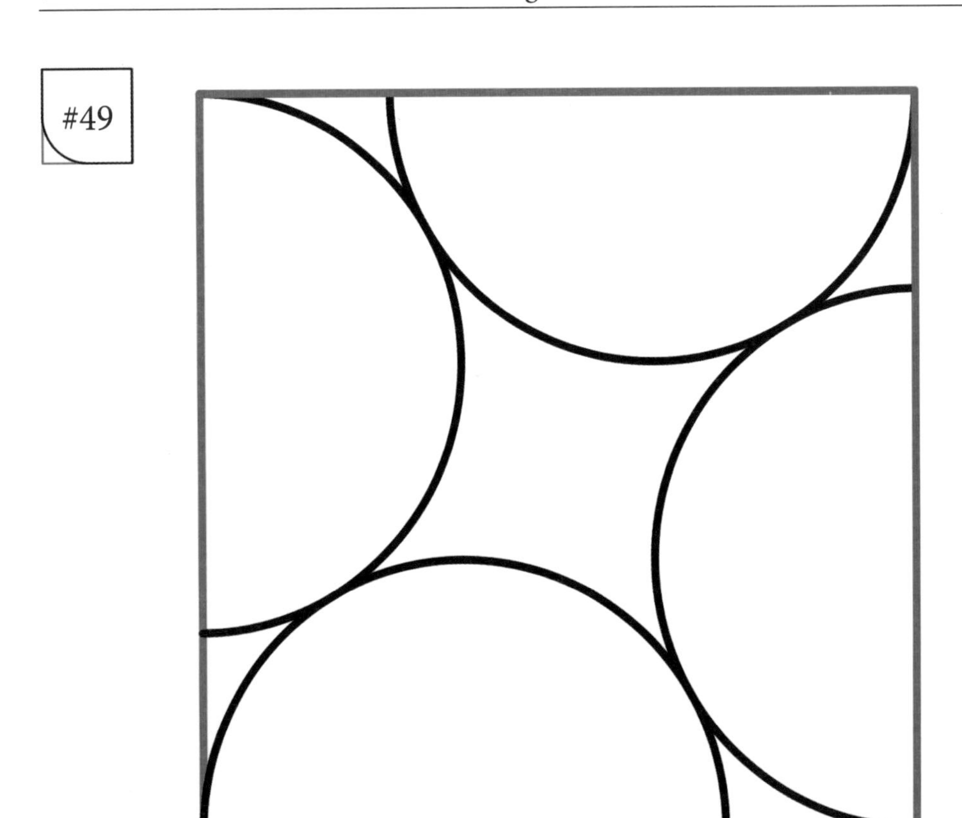

Four congruent semicircles with radius 2 cm are constructed within a square as shown. What is the area of the square?

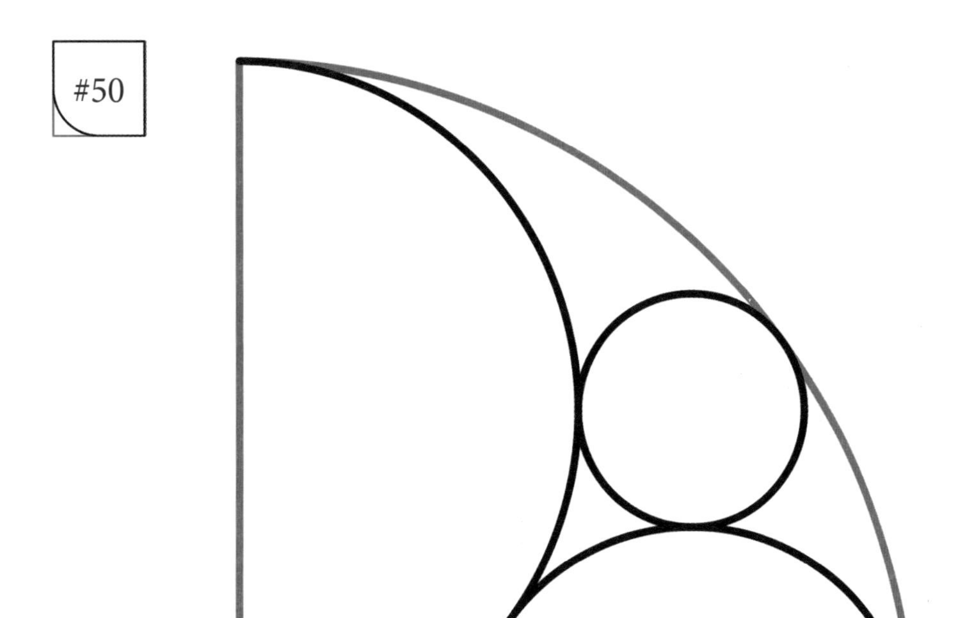

A quarter of a circle with radius 6 cm is shown with two semicircles and one circle constructed inside it such that each is tangent to the others. Find the radii of the smallest semicircle and of the small circle.

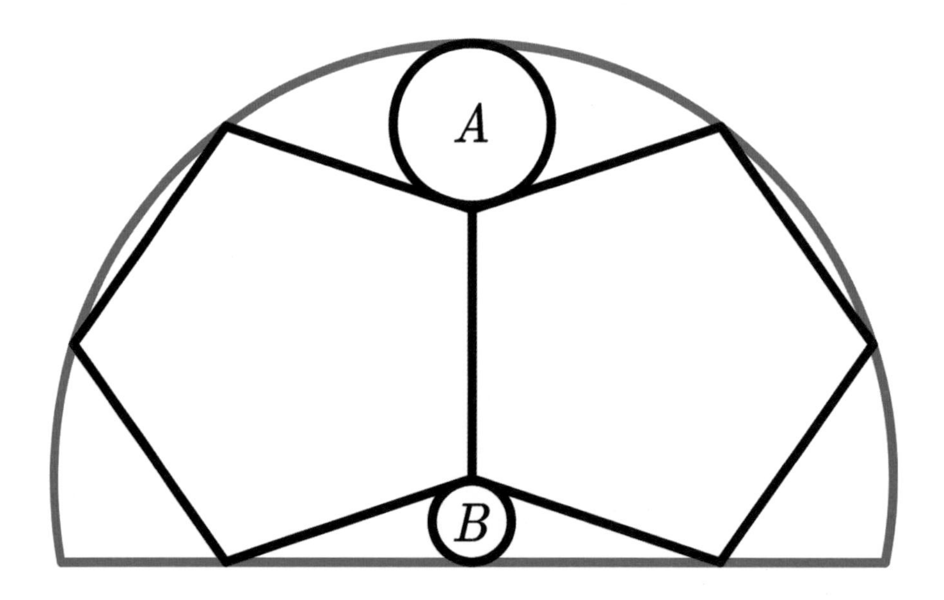

Two congruent regular pentagons share a side and are inscribed into a circular segment as shown. Two small circles are tangent to the pentagons and the segment. Show that the radius of circle *A* is double the radius of circle *B*.

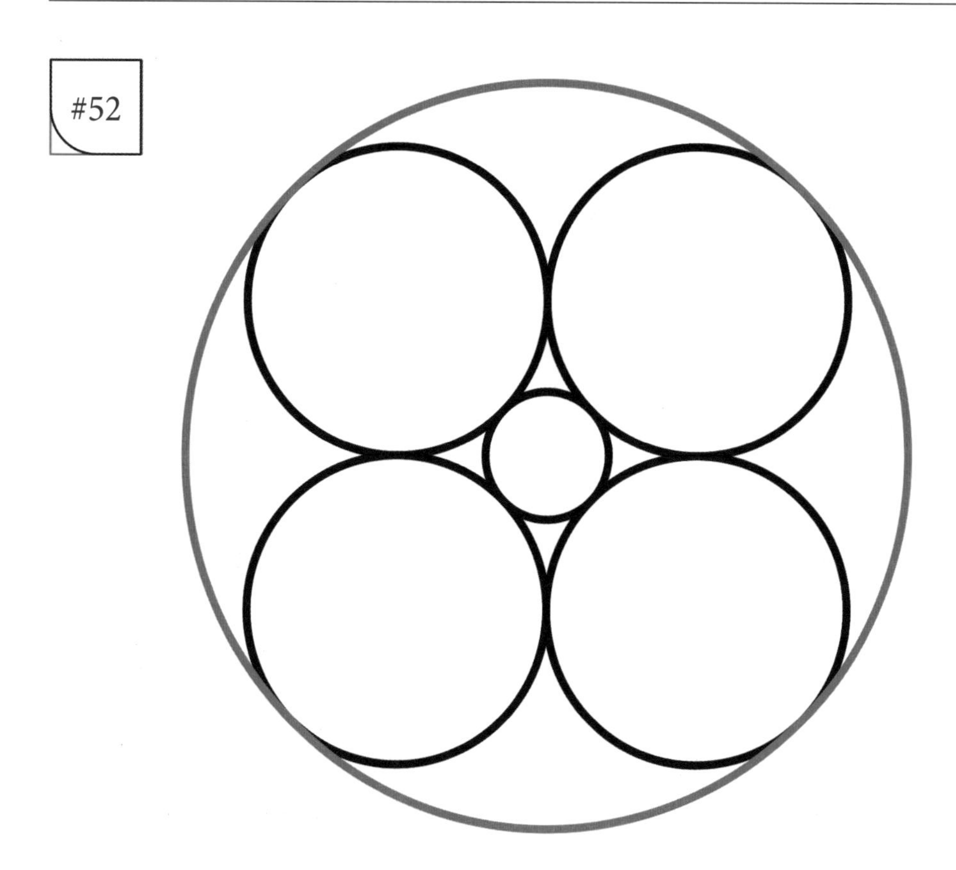

If the outer circle is a unit circle, find the radius of the smallest circle.

#53

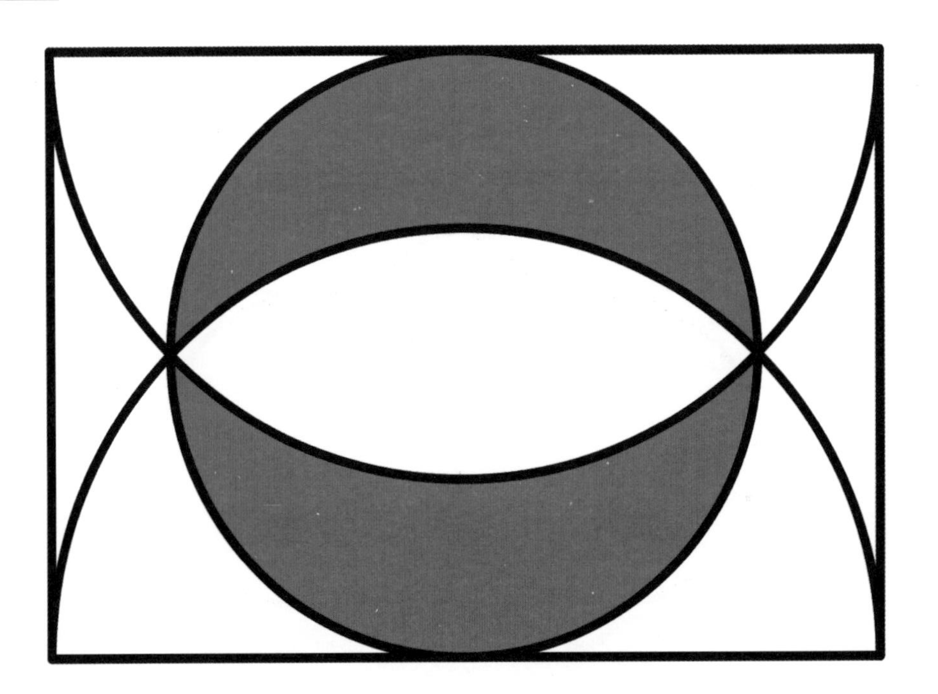

Two semicircles are constructed on the sides of a rectangle measuring 2 cm by 1 cm as shown. A circle is tangent to the longest sides of the rectangle and passes through the intersections of the semicircles. Find the combined area of the shaded lunes.

$2\sqrt{2}$ Actually.

Solutions and Answers

#49. Answer: $8(2 + \sqrt{3})$ cm^2

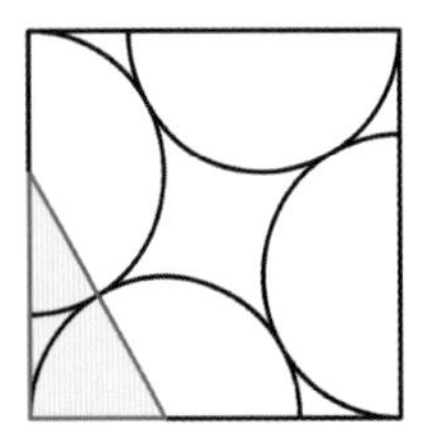

The triangle drawn has hypotenuse 4 cm and legs 2 cm and $\sqrt{12}$ cm. So the area of the square is:

$(2 + \sqrt{12})^2 = 8(2 + \sqrt{3})$ cm^2

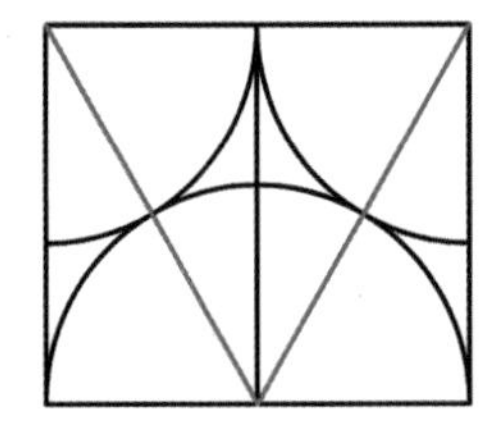

We can rearrange the highlighted triangles into a rectangle with sides 4 cm and $\sqrt{12}$ cm. So the total area of the original square is $4^2 + 4\sqrt{12} = 8(2 + \sqrt{3})$ cm^2

#50. Answer $\{R = 2$ cm, $r = 1$ cm$\}$:

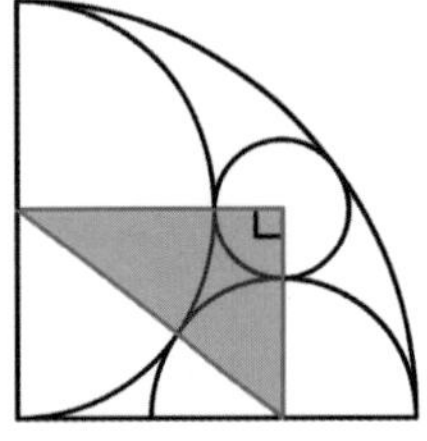

Let R be the radius of the smallest semicircle, and r that of the small circle. Then, from the diagram:

$(3 + R)^2 = 3^2 + (3 + r)^2$. Since $R + r = 3$, we get:

$(3 + R)^2 = 3^2 + (6 - R)^2 \Leftrightarrow 18R - 27 = 9 \Leftrightarrow R = 2$ cm.

Therefore the radius of the circle is $r = 3 - R = 1$ cm.

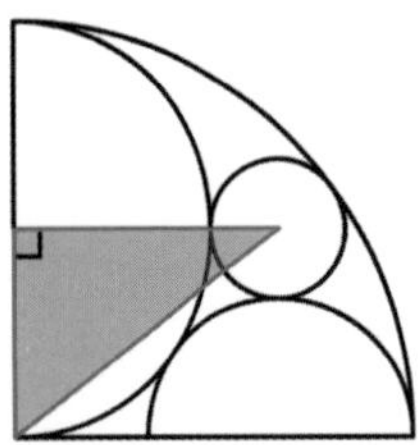

Same notations. Then from the diagram,

$(6 - r)^2 = (3 + r)^2 + 3^2 \Leftrightarrow 27 - 18r = 9 \Leftrightarrow r = 1$

Therefore the radius of the smallest semicircle is $R = 2$ cm

#51.

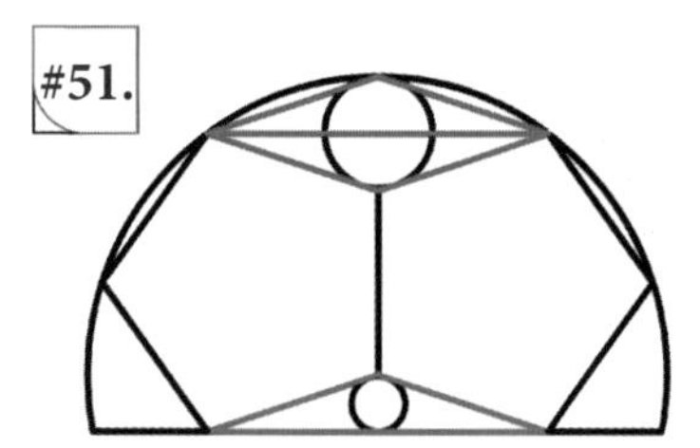

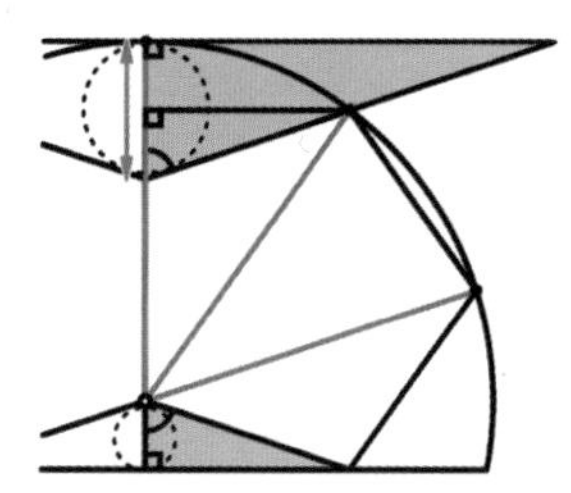

The congruent red isosceles triangles (think of their sides as both tangents from a point and sides of the pentagons) highlight that the radius of the larger circle is equal to the diameter of the smaller circle.

The marked angles measure 180 – 108 = 72°. If the pentagons have unit sides, the diagonal measures $\varphi = \frac{1+\sqrt{5}}{2}$ which is the radius of the big circle in the figure. The leg of the big shaded triangle (with the double arrow) measures $\varphi - 1 = \frac{-1+\sqrt{5}}{2} = 2\cos(72)$ so it is equal to twice the corresponding leg of the small shaded triangle. So the radius of circle A is double the radius of circle B.

#52. Answer $3 - 2\sqrt{2}$

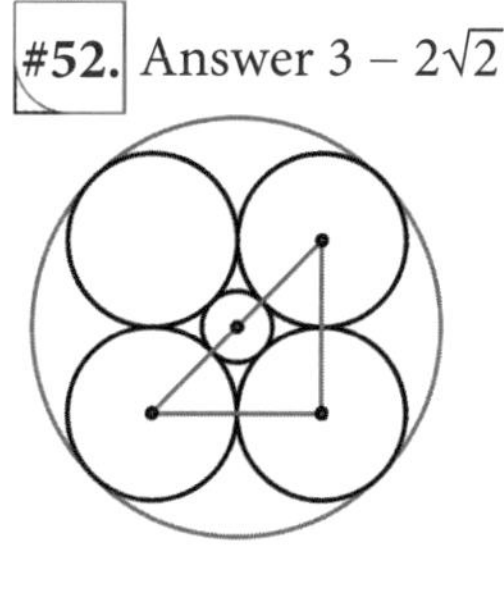

Assume that the 4 middle circles have radius 1 rather than the largest circle. The triangle in the diagram therefore would have side lengths 2, 2, $2+2r$. Using Pythagoras, we can determine the ratio of radii is $1 : (\sqrt{2} - 1)$.
Now we can take a radius of the largest circle and write it as $2 + (\sqrt{2} - 1)$, or $\sqrt{2} + 1$. Hence the ratio between all 3 circles is $(\sqrt{2} + 1) : 1 : (\sqrt{2} - 1)$. Ignoring the middle circle, if we divide both sides of the ratio by $(\sqrt{2} + 1)$ we get $1 : \frac{\sqrt{2} - 1}{\sqrt{2} + 1}$ or $1 : 3 - 2\sqrt{2}$

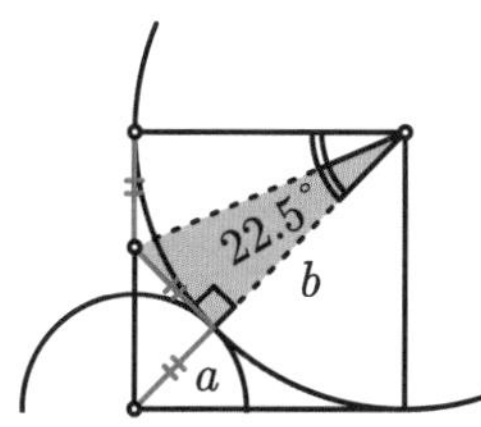

Let a and b be the small and medium radii respectively.
$\tan(45/2) = \frac{a}{b}$ so $b = \frac{a}{\tan(45/2)} = \frac{a}{\sqrt{2} - 1}$. Moreover, $a + 2b = 1$, so $a + \frac{2a}{\sqrt{2} - 1} = 1 \Leftrightarrow a(1 + 2(\sqrt{2} + 1)) = 1$ so:
$a = \frac{1}{3+2\sqrt{2})} = 3 - 2\sqrt{2}$

#53. Answer: $\frac{1}{2}$ cm^2

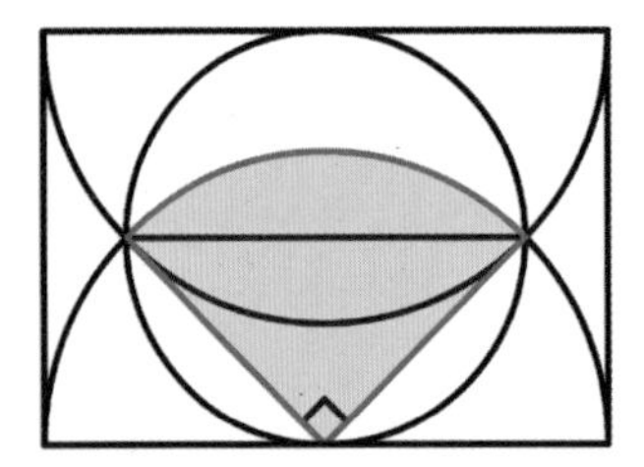

The area of the shaded sector is
$(\frac{\sqrt{2}}{2})^2 \times \frac{\pi}{4} = \frac{\pi}{8}$.
So the top segment is:
$\frac{\pi}{8} - (\frac{\sqrt{2}}{2})^2 \times \frac{1}{2} = \frac{\pi - 2}{8}$
So the lune has area $\frac{\pi}{8} - \frac{\pi-2}{8} = \frac{1}{4}$ and therefore total area is $\frac{1}{2}$ cm^2

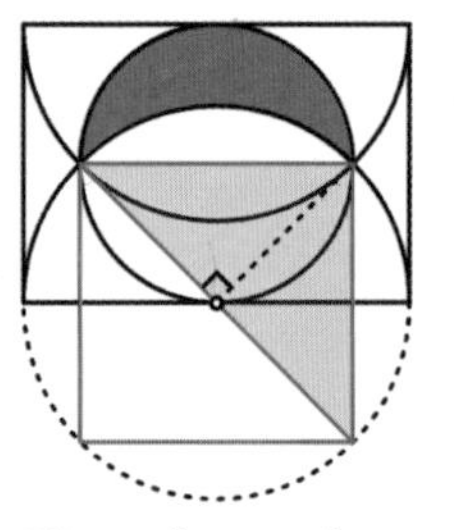

If you know the result about the lunes of Hippocrates you can notice that one red lune is on a leg of this shaded right isosceles triangle whose area is that of the two lunes:

$1^2 \div 2 = \frac{1}{2}$ cm^2